CATALYST

A framework for success

Carol Chapman

Moira Sheehan

Heinemann

Inspiring generations

Contents

T indicates Think about spread

Introduction

Welcome to Catalyst

This is the second of three books designed to help you learn all the science ideas you need during Key Stage 3. We hope you'll enjoy the books as well as learning a lot from them.

The first four pages of the book look back at what you learnt about energy, cells, particles, forces and interdependence in year 7. These are called Key ideas and the pages will help you to remember some important facts about them. You will learn more about these ideas as you go through the book.

This book has twelve units which each cover a different topic. The units have two types of pages:

Learn about:

Most of the double-page spreads in a unit introduce and explain new ideas about the topic. They start with a list of these so that you can see what you are going to learn about.

Think about:

Each unit has a double-page spread called Think about. You will work in pairs or small groups and discuss your answers to the questions. These pages will help you understand how scientists work and how ideas about science develop.

On the pages there are these symbols:

a Quick questions scattered through the pages help you check your knowledge and understanding of the ideas as you go along, for example,

> **a** **Use the particle model to explain why the liquid will not squash.**

Questions

The questions at the end of the spread help you check you understand all the important ideas.

For your notes:

These list the important ideas from the spread to help you learn, write notes and revise.

Do you remember?

These remind you of what you already know about the topic.

Did you know?

These tell you interesting or unusual things, such as the history of some science inventions and ideas.

At the back of the book:

Glossary

All the important scientific words in the text appear in bold type. They are listed with their meanings in the Glossary at the back of the book. Look there to remind yourself what they mean.

Index

There is an Index at the very back of the book, where you can find out which pages cover a particular topic.

Activities to help or check your learning:

Your teacher may give you these activities from the teacher's materials which go with the course:

Unit map

You can use this to think about what you already know about a topic. You can also use it to revise a topic before a test or exam.

Starters

When you start a lesson this is a short activity to introduce what you are going to learn about.

Activity

There are different types of activities, including investigations, that your teacher can give you to help with the topics in each spread in the pupil book.

Plenaries

At the end of a lesson your teacher may give you a short activity to summarise what you have learnt.

Homework

At the end of a lesson the teacher may give you one of the homework sheets that go with the lesson. This will help you to review and revise what you learnt in the lesson.

Pupil checklist

This is a checklist of what you should have learnt to help you with your revision.

Test yourself

You can use this quiz at the end of each unit to see what you are good at and what you might need to revise.

End of unit test Red

This helps you and your teacher check what you learnt during the unit, and measures your progress and success.

Heinemann Educational Publishers
Halley Court, Jordan Hill, Oxford OX2 8EJ
Part of Harcourt Education

Heinemann is the registered trademark of
Harcourt Education Limited

© Carol Chapman, Moira Sheehan 2003

First published 2003

07 06 05 04 03
10 9 8 7 6 5 4 3 2 1

British Library Cataloguing in Publication Data is available
from the British Library on request.

ISBN 0 435 76030 0

Edited by Diona Gregory, Ruth Holmes and Sarah Ware

Designed and typeset by Ken Vail Graphic Design

Original illustrations © Harcourt Education Limited 2003

Illustrated by Jeff Edwards, Stuart Harrison, David Lock, Richard Morris, John Plumb,
Sylvie Poggio Artists Agency (Rhiannon Powell and Lisa Smith), Simon Girling & Associates
(Mike Lacey)

Printed in the UK by Scotprint.

Picture research by Jennifer Johnson

Acknowledgements

The authors and publishers would like to thank the following for permission to use
copyright material: **5 a day leaflet p5** Reproduced with permission of the Department of
Health and the Controller of Her Majesty's Stationary Office, © Crown copyright.

The publishers have made every effort to trace the copyright holders, but if they have
inadvertently overlooked any, they will be pleased to make the necessary
arrangements at the first opportunity.

p75 The concept of 'lateral thinking' was orginated by Edward de Bono.

For photograph acknowledgements, please see page vii.

Tel: 01865 888058 www.heinemann.co.uk

The author and publishers would like to thank the following for permission to use photographs:

T = top   **B** = bottom   **L** = left   **R** = right   **M** = middle

SPL = Science Photo Library

Cover: Getty Images.

Page 2, **T**: Photofusion; 2, **M**, **B** x3: Andrew Lambert; 3: SPL/John Paul Kay, Peter Arnold ; 4, x2: Corbis; 5: SPL/Oscar Burrie; 7, **L**: Gareth Boden; 7, **R**: Holt Studios/ Nigel Cattlin; 9: Wellcome Trust, courtesy of Charles C Thomas Publishers; 10: SPL/Eye of Science; 11: SPL; 14, **T**: Empics; 14, **B**: SPL/Chris Knapton; 15, **T**: Photodisc; 15, **B**: SPL/Andrew Syred; 16: Corbis; 21: Wellcome Trust; 24: Corbis; 25, x2: Pete Morris; 26, **L**: SPL/CNRI; 26, **M**: SPL/Dr P Marazzi; 26, **R**: SPL/HC Robinson; 27: SPL/Eye of Science; 28, **T**: Pete Morris; 28, **M**: Philip Parkhouse; 28, **B**: Mary Evans Picture Library; 29, **T**: SmithKlein Beecham; 29, **B**: Pete Morris; 30, **T**: BAL; 30, **B**: Wellcome Trust; 32: John Beck; 33, **T**: Panos Pictures; 33, **B**: SPL/Noah Poritz; 34: SPL/TEK Image; 36, **T**: NHPA/James Carmichael Jr; 36, **ML**: Wildlife Matters; 36, **MR**: Garden Matters/Steffie Shields; 36, **BL**: SPL/John Howard; 36, **BR**: SPL/Alex Bartel; 37, **TL**: SPL/Claude Nuridsany & Marie Perennou; 37, **TR**: SPL/Simon Fraser; 37, **BL**: SPL/Simon Fraser; 37, **BR**: SPL/Eye of Science; 38, **T**: Garden & Wildlife Matters; 38, **M**, **B**: Peter Gould; 39, **TL**: SPL/Hermann Eisenbeiss; 39, **TR**: Oxford Scientific Films/Colin Milkins; 39, **B**: Robert Harding/Adam Woolfitt; 40: Oxford Scientific Films/GI Bernard; 44, x2: Andrew Lambert; 46, **T**; Andrew Lambert; 46, **M**: Corbis; 46, **B**: Peter Gould; 47, **L**: Panos; 47, **R**: SPL/Bernhard Edmaier; 49: Mary Evans Picture Library; 50, **T**, **BR**: Gareth Boden; 50, **BL** x3: Peter Gould; 51: Gareth Boden; 54, x4: Gareth Boden; 56: Peter Gould; 57, x5: Peter Gould; 58, x2: Peter Gould; 59: Peter Gould; 60, **T**: Pete Morris; 60, **B**: Photodisc; 61: Moira Sheehan; 63: SPL/Jerome Yeats; 66, x4: GSF Picture Library; 68, **TR**, **BR**: GSF Picture Library; 68, **TL**: Andrew Lambert; 68, **M**: Environmental Images/John Morrison; 68, **BL**: Robert Harding/Roy Rainford; 69, x2: Peter Gould; 70, **T**, **MR**: GSF Picture Library; 70, ML: Corbis; 70, **B**: Environmental Images/Clive Jones; 71: GSF Picture Library; 72: Corbis; 73: The Natural History Museum; 74, x3: The Natural History Museum; 76, **L**, **M**: GSF Picture Library; 76, **R**: Corbis; 77, **TL**, **MR**, **BR**: Corbis; 77, **TR**, **ML**, **BL**: GSF Picture Library; 78, **T** x3: GSF Picture Library; 78, **B** x2: Pete Gould; 80, **TL**, **BL**: GSF Picture Library; 80, **TR**, **BR**: Corbis; 81, **TL**: Corbis; 81, **TM**, **TR**: GSF Picture Library; 81, **B**: Digital Vision; 82, **L**, **R**: Corbis; 82, **M**: GSF Picture Library; 84, **TL**: SPL/Chris Priest & Mark Clarke; 84, **TR**, **BL**: Peter Gould; 84, **BR**: Pete Morris; 87, **TL**, **BM**: Gareth Boden; 87, **TM**: Stock Market/Kunio Swaki; 87, **TR**: Corbis; 87, **BL**: SPL/F Chillmaid; 87, **BR**: SPL/James Holmes; 88, **T**: Milepost 92$^1/_2$; 88, **B**: Corbis; 94, **T**: Pete Morris; 94, **B**: Corbis/Harcourt Index; 95, x2: Peter Gould; 96: Empics; 97, **T**: SPL; 97, **B**: Empics/Ted Leicester; 98: Peter Gould; 100, **T**: Bea Thomas; 100, **M**: Richard Thomas; 100, **B** x2: Peter Gould; 101, **T**: Gareth Boden; 101, **B**: Peter Gould; 102, **T**: National Maritime Museum; 102, **B**: ActionPlus; 103, **T**: Peter Gould; 103, **B**: Corbis; 104, x3: Peter Gould; 106, **T**: Alamy; 106, **M**, **BR**: Peter Gould; 106, **BL**: Gareth Boden; 107, **T**: Gareth Boden; 107, **B**: Milepost 92$^1/_2$; 110, **T**: Corbis; 110, **M**: Lonely Planet; 110, **B**: Andrew Lambert; 111, **TL**: Pete Morris; 111, **TR**: Alvey and Towers; 111, **M1**: SPL/Adam Hart-Davis; 111, **M2**: SPL/Space Telescope Science Institute/NASA; 111, **B**: SPL/David Nunuk; 112: Peter Gould; 113: SPL/David Scharf; 114, x2: Andrew Lambert; 115: Bruce Coleman Collection/Kim Taylor; 116, **T**: SPL/David Parker; 116, **M**, **B**: Corbis; 118: SPL/Vaughan Fleming; 120, **T**: Trevor Clifford; 120, **B**: Peter Gould; 122, **L**: Corbis/Harcourt Index; 122, **M**: SPL/Richard R Hansen; 122, **R**: FLPA/R Wilmshurst; 123: SPL/Pat & Tom Leeson; 124, **T**: SPL/Quest; 124, **B**: Bea Thomas; 126, **L** : Redferns; 126, **R**: Holt Studios; 127: SPL/R Maisonneuve, Publiphoto Diffusion; 129, **T**: SPL; 129, **M**: Gareth Boden; 129, **BL**: SPL/Tek Image; 129, **BR**: SPL/Hank Morgan.

Key ideas

You have learned about five key scientific ideas: energy, cells, particles, forces and interdependence.

Learn about:
- Energy
- Cells
- Particles
- Forces
- Interdependence

Energy is probably the most important idea in science.

Energy

Energy makes things happen.

Energy can be **transferred**. It can be transferred by light, sound, heat, electricity and movement.

Energy can be **stored**. It can be stored in materials that have been stretched or compressed, like bows and springs and squashed rubber.

Light energy, sound energy, thermal (heat) energy and electrical energy are all energy on the move.

Strain energy.

Energy can be **stored** in chemicals, like a fuel and oxygen before burning.

Energy can be stored in things that have been lifted up, like a skier at the top of a slope.

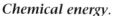

Chemical energy.

Gravitational energy.

*We measure energy in **joules, J**, or **kilojoules, kJ**.*

1000 J = 1 kJ.

Cells

All living things are made of **cells**.

Animal and plant cells contain a **nucleus, cytoplasm** and a **cell membrane**.

Plant cells also have a **cell wall**. Some plant cells have a large **vacuole** filled with liquid. Plant cells from the green parts of the plant also have **chloroplasts**.

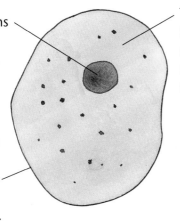

The nucleus contains the instructions to make the cell work.

The cytoplasm is where chemical reactions like respiration take place.

The cell membrane lets some things in and out of the cell, but other things can't go through it.

cell membrane

The vacuole keeps the cell membrane pressed against the cell wall.

The cell wall is like a box.

nucleus

The chloroplasts contain **chlorophyll**, which traps light energy.

In multicellular plants and animals, similar cells work together in a **tissue** and tissues are grouped together to make **organs**.

The heart is an example of an organ. It contains muscle tissue, nervous tissue and blood vessels.

Viruses are not made of cells, but no one is sure if they are alive!

*Scientists use their ideas about particles, or the **particle model**, to explain why we have solids, liquids and gases.*

Particles

All materials are made up of **particles**.

The particle model

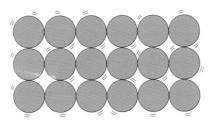

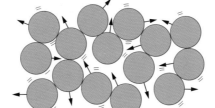

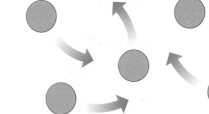

In solids the particles:

- are in rows
- are touching
- vibrate on the spot
- are held together.

In liquids the particles:

- are disordered
- are touching
- vibrate and slide over each other
- are held together.

In gases the particles:

- are disordered
- are far apart
- zoom about
- are not held together.

The particle model can also be used to explain **changes of state**, like melting, freezing, boiling and condensing.

It can also be used to explain dissolving, expanding and diffusion.

Key ideas (continued)

Forces

Forces are pulls, pushes and twists. Forces have both size and direction. There is usually more than one force on an object. You have to think about the size and the direction of each force to decide what will happen.

Balanced forces

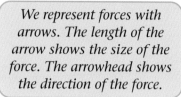

We represent forces with arrows. The length of the arrow shows the size of the force. The arrowhead shows the direction of the force.

The boy does not move because the forces are equal but opposite.

Unbalanced forces

The boy moves to the right.

The boy moves to the left.

We give forces names to tell us what is causing them. **Weight** is the force we feel because of gravity. Other forces include **friction**, **upthrust** and **air resistance**.

*We measure forces in **newtons, N.***

Interdependence

Each organism relies on other organisms to stay alive.

The green hairstreak butterfly

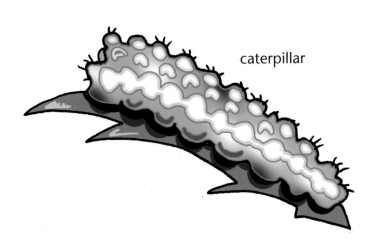
caterpillar

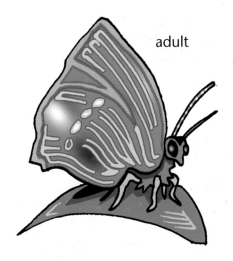
adult

Birds and other predators eat the green hairstreak caterpillars and butterflies.

The green hairstreak butterfly pollinates plants when it goes from flower to flower collecting nectar.

It needs plants like bilberry, gorse and heather for:

- food as a caterpillar (leaves)
- food as an adult (nectar)
- camouflage
- shelter
- a place to lay its eggs.

If they had no predators, the green hairstreak caterpillars would eat all the leaves and the plants would die. Then the caterpillars and butterflies would starve.

We show how organisms rely on other organisms for their food using **food chains** and **food webs**.

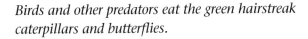

bilberry → green hairstreak → bird

- The **producer** always comes first.
  It makes its own food using sunlight.
- **Consumers** eat other organisms.
- The arrows show the flow of energy.

The birds and other predators need the green hairstreak for food. But without the predators all the green hairstreaks would run out of food and die.

The green hairstreak needs the plants, and the plants need the green hairstreak.

A food web is many food chains put together.

A1 What's in food?

Food for thought ...

All animals and plants need food. Food gives us the energy we need to carry out our life processes. Plants make their own food by photosynthesis. Animals have to find their food. It is easy for us to go to the nearest supermarket. But do we really need all this food? Have you ever wondered where it goes inside you, and what happens to it?

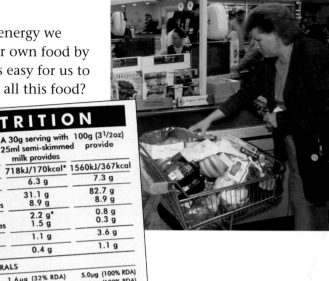

What are nutrients?

Food contains many different substances. The useful substances that food contains are called **nutrients**.

ⓐ Look at this food label. What are the main types of nutrient in this food?

The main nutrients are **carbohydrates**, **fats** and **proteins**. They make up most of the food we eat.

NUTRITION		
TYPICAL COMPOSITION	A 30g serving with 125ml semi-skimmed milk provides	100g (3½oz) provide
Energy	718kJ/170kcal*	1560kJ/367kcal
Protein	6.3 g	7.3 g
Carbohydrate	31.1 g	82.7 g
of which sugars	8.9 g	8.9 g
Fat	2.2 g*	0.8 g
of which saturates	1.5 g	0.3 g
Fibre**	1.1 g	3.6 g
Sodium	0.4 g	1.1 g
VITAMINS/MINERALS		
Vitamin D	1.6µg (32% RDA)	5.0µg (100% RDA)
Thiamin	0.5mg (34% RDA)	1.4mg (100% RDA)
Riboflavin	0.7mg (44% RDA)	1.6mg (100% RDA)
Niacin	6.5mg (36% RDA)	18.0mg (100% RDA)
Vitamin B6	0.7mg (34% RDA)	2.0mg (100% RDA)
Folic acid	127.5µg (63% RDA)	400.0µg (200% RDA)
Vitamin B12	0.8µg (80% RDA)	1.0µg (100% RDA)
Pantothenic acid	2.2mg (37% RDA)	6.0mg (100% RDA)
Iron	4.3mg (30% RDA)	14.0mg (100% RDA)
RDA = Recommended Daily Allowance		

This pack contains approx 16 servings

INFORMATION

*Calories/Fat per serving with whole milk: 195 cals/5g

Carbohydrates give you energy. They are found in bread, potatoes, cakes and sweets. Sugar and starch are both carbohydrates.

Proteins help your body to grow and repair itself. They are found in meat, fish, eggs, peas, beans and milk.

Fats also give you energy. You have a layer of fat under your skin that keeps you warm. Fats are found in butter, margarine, full-fat milk and meat.

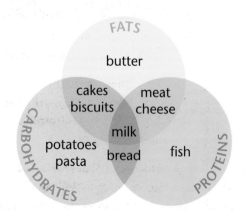

FATS

butter

cakes biscuits

meat cheese

CARBOHYDRATES

milk

potatoes pasta

bread

fish

PROTEINS

More than one nutrient

Jane and Tony tested a range of foods to see if they contained carbohydrate, fat and protein. They found that some foods contained two nutrients and others contained three. Their results are shown in the diagram on the left.

ⓑ (i) Which foods contain carbohydrate and fat?
(ii) Which foods would you recommend to someone who had been ill and needed to repair their body? Explain your answer.

Small amounts – vitamins and minerals

Vitamins are needed in very small amounts to keep the body healthy. Different vitamins have different functions. If a vitamin is missing from someone's diet, it can cause a disease. We say that the person is **deficient** in that vitamin.

Vitamin C is found in fruit and green vegetables. Sailors on long voyages used to become deficient in vitamin C because they did not eat any fresh food. They got the disease **scurvy**. Their gums bled and their teeth fell out. If they cut themselves, the skin did not heal.

Vitamin D is found in milk and butter. It is also made by your body in sunlight. Vitamin D gives you strong bones and teeth. If this vitamin is deficient it can cause a disease called **rickets** where the bones are soft.

Your body also needs **minerals** in small amounts. Different minerals have different functions.

- **Calcium** is a mineral found in milk and cheese. You need calcium for healthy teeth and bones.

- **Iron** is found in liver and eggs and is used to make blood. If iron is missing from the diet it can cause anaemia.

- **Iodine** is found in foods such as fish. The thyroid gland in the neck uses iodine to produce a hormone that makes you grow.

Fibre and water

You also need **fibre** in your diet. Fibre is sometimes called **roughage** and it is found in cereals, fruit and vegetables. It helps food to keep moving through the gut. Fibre prevents you from becoming **constipated** and may reduce the risk of some types of cancer.

C Fibre comes from plants. Think about the differences between animal cells and plant cells. What do you think gives plants this fibre?

You take in **water** when you eat and drink. You would die in a few days without water. All the chemical reactions in your body take place in water.

Did you know?

People who do not have enough iodine can get a disease called **goitre**. The thyroid gland becomes very large.

Did you know?

Your body loses about two litres of water a day, most of it as urine.

Questions

1 Explain the meaning of the following words. Give examples of each.

   a nutrient   b deficiency.

2 Which foods mentioned on these two pages would contain the following nutrients:

   a protein, fat and calcium?   b protein and fibre?
   c protein and iron?

3 Samina said water is not a nutrient. What do you think?

4 Design a quiz sheet for other pupils in your class about all the different nutrients in the human diet. Include questions and answers on your sheet.

For your notes:

- Food contains useful substances called **nutrients**.

- The main types of nutrient in food are **carbohydrates, fats, proteins, vitamins** and **minerals**.

- **Fibre** and **water** are also needed for a healthy diet.

Getting the balance right

Your body needs nutrients for energy and to keep it working properly so you stay fit and healthy. A diet that has the right amounts of all the nutrients is called a **balanced diet**. A balanced diet can be different for different people. The foods which can make it up can be different.

How much should we eat?

Government scientists advise us how much of each nutrient we should eat a day. They set **recommended daily intakes (RDI)** or **recommended daily allowances (RDA)** for some of the nutrients and energy that make up a balanced diet.

The table below shows some RDAs for different people.

Person (age in years)	Energy in kJ	Protein in g	Vitamin C in mg	Iron in mg
male 10–12	10 900	30	20	5–10
female 10–12	9800	29	20	5–10
adult male	12 600	37	30	5–9
adult female	9200	29	30	14–28

a Describe any differences between:
   (i) males and females   (ii) children and adults.

b (i) What differences can you see in the RDAs for iron?
   (ii) Can you explain this?

Balancing the energy

The amount of energy you need in your food depends on how much energy your body uses up every day. This depends on whether you are growing, how active you are and the size of your body. Men need more energy than women.

A teenage boy uses about 12 200 kJ of energy every day, while a girl of the same age uses about 9600 kJ. A person doing a very active job such as a builder uses a lot more energy than an office worker of the same age and sex.

Sports people use a lot of energy and they need to take in energy very quickly. They often have drinks that contain a lot of sugar such as glucose that can be used for energy very quickly.

It is important to take in the right amount of energy in your food. Some people take in more energy than they use up. They risk becoming fat. People who are very overweight for their height are described as **obese**.

Obese people may eat less in order to lose weight. You should always speak to your doctor before going on a diet. Sometimes people can suffer from eating disorders such as **anorexia nervosa**. They do not eat enough and they become very underweight. This disease can kill.

Diet in pregnancy

Pregnant women are advised to supplement their diet with folic acid. This helps prevent certain kinds of disability in the baby. Women also need extra nutrients in the last three months of pregnancy. They may take extra calcium for the growing bones of the fetus, protein for its muscles and cells and iron for its blood.

c What do you think 'supplement' means?

Different diets

Vegetarians do not eat any meat. To give them the protein they need in a balanced diet they eat nuts, seeds and cereals. Balanced vegetarian diets can be very healthy as they contain less fat and more fibre than a diet including meat.

People from different cultures eat different foods. This may be due to religion, tradition or simply what is available in the country. In many Asian countries, such as China, India and Pakistan, the main food is rice. In some countries in Africa the main food is maize.

In developed countries, such as the UK, the diet often contains too much fat and salt. A person who eats a high-fat diet is more likely to have a heart attack. A diet high in salt can increase the risk of having strokes. This kind of diet is often termed the 'Western diet'.

We also eat too little fresh fruit and vegetables. Government scientists encourage us to eat five portions of fruit or vegetables a day to keep healthy.

Questions

1 Jake is a vegetarian. His friend Zoe enjoys eating burgers and chips. Jake thinks that his diet is probably more healthy than Zoe's. Explain why he may be correct.

2 There are no RDIs or RDAs for carbohydrate or fat in Western countries.

   a Why do you think this is the case?

   b Why are carbohydrates and fats important for the diet?

3 Decide whether you think each of these statements is true or false. Explain your decisions.

   a An office worker uses less energy than a builder.

   b Vegetarians are at risk of eating too little roughage.

   c Athletes need lots of glucose when they compete.

For your notes:

- A diet that has the right amount of each nutrient is called a **balanced diet**.

- It is important to balance the energy in your food with the energy your body uses.

Breakfast on the move

Have you ever missed breakfast and had a cereal bar on the way to school instead? Cereal bars are advertised as the modern alternative to breakfast. They contain lots of nutrients including vitamins and fibre and they are easy to eat on the move – particularly useful if you are running late!

But some experts in healthy eating are worried that cereal bars are loaded with salt, sugar and fats.

a Why are cereal bars a popular alternative to breakfast?

A salty taste?

The average person takes in 9 g of salt each day, but some scientists recommend that we should reduce this to 6 g. They think that too much salt in the diet can cause high blood pressure, which can increase the risk of heart disease and strokes.

Other scientists disagree. They say that although cutting down your salt intake reduces high blood pressure, this does not mean that too much salt raises your blood pressure.

A sweet tooth?

When we eat sugary food, microorganisms called bacteria on our teeth feed on the sugar. This produces the acid that causes tooth decay.

Too much fat?

Eating too much fat, especially animal fats, can increase the risk of heart disease. A layer of fat can build up in your blood vessels. This makes the blood vessels narrower, slowing down blood flow, sometimes stopping it. This is especially dangerous in the blood vessels:

- that take blood to the heart, where it causes a heart attack.

- that take blood to the brain, where it causes a stroke.

> **Did you know?**
>
> The sugar in a cereal bar is more likely to stick to your teeth and cause tooth decay than the sugar in a bowl of breakfast cereal with milk. The milk washes the sugar away from your teeth!

NUTRITIONAL INFORMATION
100 g provides:
Energy 1900 kJ/450 kcal.
Protein 9 g.
Carbohydrates 67 g (of which are sugars 37 g, starch 30 g).
Fat 16 g (of which saturates 14 g).
Fibre 1.6 g.
Salt 1.6 g.
Vitamins: Thiamine (B$_1$) 0.9 mg (65%). Riboflavin (B$_2$) 1.0 mg (65%). Niacin 11.3 mg (65%). Vitamin B$_6$ 1.3 mg (65%). Folic acid 125 µg (65%).
Minerals: Calcium 720 mg (90%). Iron 8.8 mg (65%).
(%) = % Recommended Daily Allowance

One Cerius bar weighs 25 g.

Some food labels tell you how much sodium is in 100 g of the food, instead of how much salt. This can be misleading because sodium and salt are not the same thing. Salt is sodium chloride. You have to multiply the amount of sodium by 2.5 to find out the actual salt content. Experts are asking for clearer labelling that says how much salt there is in 100 g of the food, as on the Cerius Bar label shown on the left.

b Look at the Cerius bar label. Do you think it is a healthy or unhealthy breakfast? Use the ingredients to justify your argument.

Passing through

Some undigested food, mainly fibre, is left in the small intestine. Fibre is made of molecules called **cellulose** that come from plant cell walls. Humans cannot digest cellulose because they cannot make the enzyme **cellulase** that is needed to break it down. Herbivores like rabbits have bacteria in their gut that produce cellulase.

In humans, undigested food such as cellulose passes into the **large intestine**. It is stored here, and water is absorbed from it into the blood to be used by the body. Later the waste is pushed out of the body as faeces through the **anus**. This process is called **egestion**.

C How are herbivores like rabbits able to digest plants?

Problems ...

If your digestive system is working well, the waste from your food should leave your body after about 36 hours. A lack of fibre in your diet can cause constipation. The faeces are stored for too long. They become dry and hard. The best way to avoid constipation is to include more fibre in your diet – eat plenty of fruit and vegetables.

An infection of the large intestine can cause **diarrhoea**. The waste is not stored for long enough and not enough water is absorbed back into the blood. A person suffering from diarrhoea should drink lots of fluids.

Did you know?

Rabbits eat their own **faeces**! When they eat grass, they produce lots of soft faeces. They eat these and digest them again before producing dry rabbit droppings.

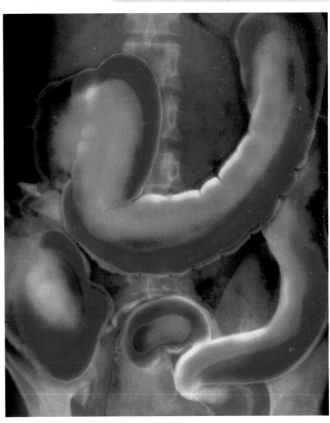

This X-ray shows the large intestine.

Did you know?

Babies can die from diarrhoea because they lose too much water from their body.

Questions

1 Write a paragraph to explain absorption using the words below.

   glucose   molecules   small intestine   starch   villi

2 The small intestine is one of the longest parts of the gut. It is folded up inside the body. Why do you think it is so long and folded?

3 Explain why vitamins and minerals can pass through the wall of the small intestine but fibre cannot.

4 Describe what happens in the large intestine. You could show this as a diagram.

5 What causes:

   a constipation?   **b** diarrhoea?
   How can they be prevented?

6 Write a poem about the journey of the food through the gut.

For your notes:

- Only small molecules can pass through the wall of the small intestine into the blood.

- Small molecules of digested food are **absorbed** into the blood.

A6 Chewing it over

Digestion of starch

Wendy took a bite of her baguette. She thought, 'The bigger the bite, the more starch there is to be digested, and the longer it takes ...'

Wendy wanted to find out whether adding a bigger amount of an enzyme to starch speeds up digestion. Starch is broken down (digested) into glucose by an enzyme called salivary amylase, which is produced in your mouth.

She used the same amount of starch each time, but added different amounts of the enzyme. She used iodine indicator to show when the starch had disappeared. The iodine indicator loses its blue-black colour when all the starch has gone. She timed how long it took for the starch to disappear each time. When the starch had disappeared, she knew it had all been digested.

Her results are shown in this table. The concentration tells you the amount of salivary amylase enzyme she added. For example, $1\,cm^3$ of 2% enzyme contains twice as much enzyme as $1\,cm^3$ of 1% enzyme.

Amount (concentration) of salivary amylase in %	Time taken to digest all the starch in seconds
1	240
2	210
3	160
4	115
5	80

Wendy wanted to see if there was a pattern in her results, so she drew a line graph. Line graphs help you see the relationship between variables in an experiment.

Wendy put the input variable (the amount of enzyme used) along the bottom, on the x-axis. She put the outcome variable (the time taken to digest the starch) up the side, on the y-axis. She plotted her results on the graph.

a Draw axes like Wendy did and plot her results.

b Can you draw a straight line through all the crosses?

Lines of best fit

In many experiments, you cannot draw a line that passes through all the points. You have to draw a line that fits most of them. This is called the **line of best fit**.

c Why do you think that lines on graphs do not always go through all the points?

Wendy's graph is shown on the right, with her line of best fit.

d What pattern do you see in the results?

e Describe the relationship between the input variable and the outcome variable.

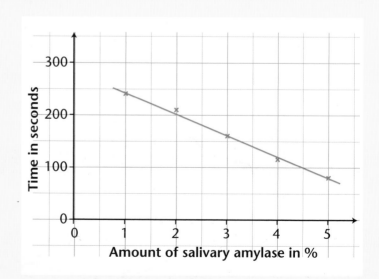

Interpreting graphs

Drawing a line of best fit helps you see the **trend** or general pattern of the results. Then you can see the relationship between the input variable and the outcome variable.

Wendy used her graph to predict how long it would probably take for other concentrations of amylase to digest starch. She drew a line up to the graph from 1.5% on the x-axis, and read off the time on the y-axis.

f Do this on your copy of Wendy's graph. What is the time on the y-axis?

g Why do you think she would want to do this?

She can also extend the line and predict how fast starch will be digested by a 6.5% concentration of amylase.

h Do this on your copy of Wendy's graph. How long will it take for the starch to be digested by a 6.5% concentration of amylase?

Other relationships

Charlotte decided to look at the way pH affects how an enzyme works. She used the same concentration of salivary amylase for all her tests, but she changed the pH of the starch solution by adding acid or alkali. She recorded the time it took for the starch to be digested. Her results are shown in this table.

i Use the data in the table to plot a graph. Try to draw a curve of best fit to join the points.

j Describe the relationship shown by your graph in terms of how the time taken to digest the starch changes with pH.

The relationship shown by the graph is not a simple relationship. To begin with, as the pH increases, the time taken to digest the starch is shorter. After a certain point, the reaction starts to take longer again.

k What pH do you think will give the shortest time for amylase to digest starch? Explain your answer.

pH	Time taken to digest all the starch in seconds
2	starch not digested
3	480
4	240
5	180
6	160
7	180
8	240
9	480
10	starch not digested

Questions

1 Why do we use lines of best fit?

2 What is the difference between the relationships shown by straight-line graphs and those shown by curved-line graphs?

3 Sometimes the results do not all fit the pattern of the graph. Think about any experiments you have carried out in the past where the results were not what you expected. Why did this happen?

B1 Food for energy

Food and energy

Energy from the food you eat keeps your muscles moving and organs working. Energy is also used for growing and repairing cells. You need energy all the time to keep your body's life processes going, even when you are asleep. Food is like a fuel for the body.

Do you remember?

Energy comes from carbohydrates and is stored as fats. Carbohydrates, fats and proteins provide your body with most of the things you need to stay healthy.

Fuels and respiration

You have to burn fuels like petrol and gas to get energy from them. Fuels react with oxygen to produce carbon dioxide and water, and energy is released. This is called **combustion**.

Your body uses a similar reaction to get the energy from food. It is called **respiration**. The fuel in respiration is glucose. The glucose comes from the digestion of food. The glucose does not burn with flames like fuels do. Respiration happens at a much lower temperature, and at a slower and more controlled rate.

Respiration can be shown in a word equation:

glucose + oxygen → carbon dioxide + water   **energy released**

ⓐ What are the reactants and products in respiration?

Where does respiration happen?

Respiration takes place in all living cells. It happens in small structures called mitochondria inside the cells. Some cells, such as muscle cells, need lots of energy. They have lots of mitochondria to carry out respiration and supply the energy.

Working out

When you exercise hard your muscles need to release more energy from glucose, so your muscle cells respire more. This uses up a lot of glucose and oxygen. Your heart beats faster and you breathe more deeply to increase the supply of glucose and oxygen to the cells.

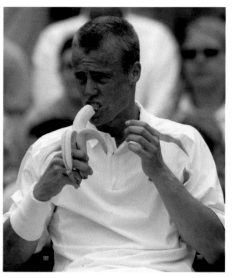

The energy in the banana helps the tennis player to keep playing.

Gas is being burnt at this onshore gas rig. It reacts with the oxygen in the air in a combustion reaction.

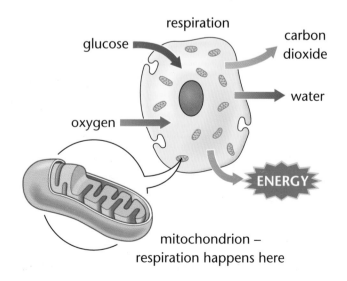

respiration

glucose → carbon dioxide

oxygen → water

ENERGY

mitochondrion – respiration happens here

The chemical energy from the glucose is mainly transferred as movement energy in the muscles, but some of it is released as heat energy. It makes you feel hot.

Go for the burn

Most of the respiration in your cells needs oxygen and is called **aerobic respiration**. But sometimes your body just can't supply enough oxygen to your cells, so they cannot carry out aerobic respiration. Then, for a short while, your cells can respire without oxygen. This happens when you run very fast for a long time.

This is called **anaerobic respiration**. It does not need oxygen. We can write a word equation for anaerobic respiration.

glucose → lactic acid   **energy released**

In animals, anaerobic respiration produces **lactic acid**. Your cells cannot respire anaerobically for very long because lactic acid is poisonous. It causes pain and cramp in the muscles. At the end of a race a sprinter pants deeply to take in lots of oxygen to break down the lactic acid.

b **When do we respire anaerobically?**

The microorganism yeast respires anaerobically to get energy. This produces ethanol (alcohol). The process is called **fermentation**. We can write a word equation for fermentation:

glucose → carbon dioxide + ethanol   **energy released**

Yeast is used in breadmaking. The carbon dioxide makes the dough rise. It is also used for making alcoholic drinks such as beer and wine.

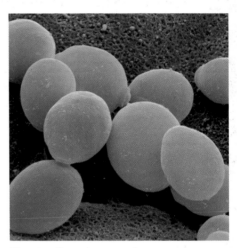

Yeast cells get energy by anaerobic respiration.

Questions

1 Why is respiration so important to all organisms?

2 How is respiration similar to burning wood?

3 All the windows in the science block are jammed shut. One small room in the block is full of people for several hours. A similar room next door has just one person in it. What do you think would be different about the air in the two rooms?

4 Explain the differences between aerobic and anaerobic respiration.

5 Explain why anaerobic respiration in yeast is a useful reaction.

For your notes:

● **Aerobic respiration** is the process by which we get energy from food.

● Aerobic respiration can be summarised by the equation:

  glucose + oxygen → carbon dioxide + water

● **Anaerobic respiration** occurs without oxygen.

Energetic forests

A forest seems a very peaceful place. In fact, all the trees are very busy making their own food. This food is called glucose.

Like humans, plants need to release the chemical energy from food by aerobic respiration. It happens in all plant cells, just as it does in all animal cells. Glucose and oxygen react, producing carbon dioxide and water, and releasing energy.

Do you remember?

Like animals, plants need energy to carry out their life processes. Plants make their own food from air and water by using light energy.

a Where in a plant does respiration take place?

Pea seeds

Miss Dupont decided to demonstrate to her pupils that plants respire. She set up an experiment with pea seeds just starting to grow. She told her pupils that the cells in the pea seeds respire to get energy for growth from the food stored in the seeds.

She wanted to show that the respiring seeds produce carbon dioxide . She used an indicator called **hydrogencarbonate indicator**. It is very sensitive and can detect a very small increase in the amount of carbon dioxide dissolved in it. It changes from red to yellow. The diagram shows the apparatus.

The indicator changed colour from red to yellow in tube 1 but not in tube 2.

b What did the pupils conclude from the experiment?

c Why did Miss Dupont set up tube 2?

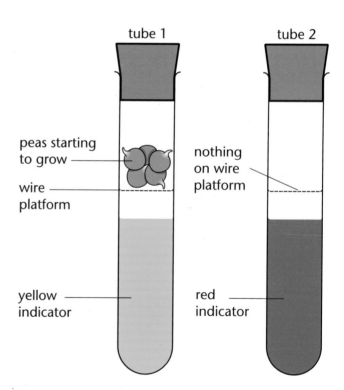

tube 1   tube 2

peas starting to grow

nothing on wire platform

wire platform

yellow indicator

red indicator

Further experiments

Some of Miss Dupont's pupils began to ask further questions about respiration.

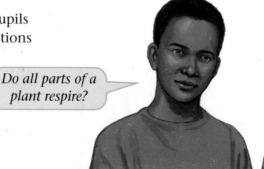

Do all parts of a plant respire?

Do small living things respire in the same way as humans and plants?

They set up an experiment to find out whether maggots and plant roots respire. They used the hydrogencarbonate indicator again. The diagram shows their apparatus after 4 hours.

d Why did the indicator change colour in tubes 1 and 2?

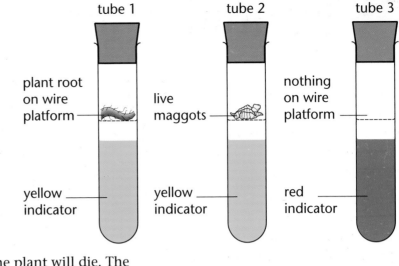

tube 1   tube 2   tube 3

plant root on wire platform

live maggots

nothing on wire platform

yellow indicator   yellow indicator   red indicator

The root of the matter

Root cells need oxygen from the air for respiration too. If the soil is waterlogged, all the air spaces in the soil are filled with water. The roots cannot get enough oxygen and so the plant will die. The glucose for respiration in root cells is made in parts of the plant that are above the ground and is transported through the stem to the root cells.

At the end of the lesson, Miss Dupont's pupils concluded that respiration takes place in every cell of all living things.

Energy release

Another class extended the enquiry. They investigated which organism respired most in 24-hours. Some of the energy released during respiration is transferred to the surroundings as heat energy. They measured the heat energy produced by maggots and germinating peas.

They set up their experiment as shown in the diagram.

A   B   C

thermometer   damp cotton wool bung   vacuum flask

maggots   germinating peas   glass beads

Their results are shown in the table.

e **(i)** Why did they set up flask C?
(ii) Why do you think they used vacuum flasks for their experiment?
(iii) Which organism respired most? Suggest a reason for your answer.
(iv) Do you think their method could be improved in any way?
(v) Is one set of results enough to base a conclusion on?

	Start temp in °C	End temp in °C
Flask A	22	32
Flask B	22	27
Flask C	22	23

For your notes:

- Plants and other organisms release energy from food by respiration, in the same way as humans do.

- Respiration takes place in every cell of all living things.

Questions

1 Which gas do plants:

a use for respiration?   **b** produce by respiration?

2 Explain why plants can die in waterlogged soil.

3 Describe a method of detecting the carbon dioxide released by cells.

Getting to the cells

All organisms need oxygen for respiration. For humans and other organisms that live on land, the oxygen comes from the air. But it needs to get inside the organisms to all the cells. Also, all cells produce carbon dioxide when they respire, and this needs to get out of the organism into the air.

In plants and some very small animals, most cells are in contact with the air. The oxygen can pass straight through the very thin cell walls and into the cells. Carbon dioxide can pass out through the cell walls into the air.

In humans and other large animals, most cells are deep inside the body. They are not in contact with the air. We need a special way of getting oxygen into the cells for respiration, and carbon dioxide out. To do this we **breathe** air in and out of lungs inside our bodies.

ⓐ **Where do all organisms get the oxygen they need for respiration?**

air can get to all the cells in the leaf

How do your lungs work?

If you put your hands on your chest, you can feel it move up and down as you breathe in and out. Your lungs are in your chest. The lungs are organs. Your lungs and tubes that take gases in and out of the lungs are an organ system called the **respiratory system**.

The air we breathe

The table shows how much oxygen and carbon dioxide there is in the air you breathe in and the air you breathe out.

Gas	Air breathed in	Air breathed out
carbon dioxide	0.04%	4%
oxygen	21%	15%
nitrogen	78%	78%
water vapour	variable	variable but a lot more

ⓑ **Explain how the content of the air you breathe in and out changes. Use the table to help you.**

Gas exchange

In the lungs, oxygen moves into the blood and carbon dioxide moves out of the blood. This is called **gas exchange**.

ⓒ **What do you think happens to the rate of gas exchange during exercise?**

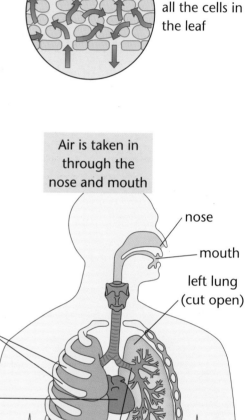

Air is taken in through the nose and mouth

nose

mouth

left lung (cut open)

ribs

heart

right lung

left bronchus

Air reaches the tiny air sacs called alveoli

Do you remember?

Gas particles move from one place to another by **diffusion**. They move from an area where there are a lot of particles to an area of where there are fewer particles, until they are evenly spread out.

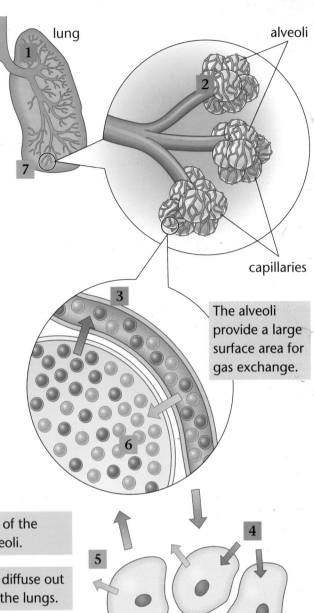

The alveoli provide a large surface area for gas exchange.

cells in the body

Oxygen in

This is how the oxygen you need for respiration gets from the air you breathe to the cells in your body.

1 Oxygen enters your body through your lungs when you breathe in air.

2 Inside the lungs are many tiny air sacs called alveoli with walls only one cell thick. Oxygen particles can diffuse through the alveoli walls quickly because they are so thin. There is a lot of oxygen in the lungs, so oxygen particles diffuse out of the lungs through the alveoli.

3 The alveoli are surrounded by many tiny blood vessels, also with very thin walls. Oxygen particles pass through the walls of the alveoli into the blood. The blood vessels carry oxygen to all the cells in the body.

4 Oxygen particles pass through the thin walls of the blood vessels into the cells. The cells respire using the oxygen and produce carbon dioxide.

Carbon dioxide out

Carbon dioxide is a waste product of respiration. This is how the cells get rid of carbon dioxide.

5 Carbon dioxide particles move from the cells across the walls of the blood vessels into the blood. They are carried back to the alveoli.

6 There are a lot of carbon dioxide particles in the blood. They diffuse out of the blood vessels through the walls of the alveoli and into the lungs.

7 The carbon dioxide leaves your lungs in the air you breathe out.

Questions

1 a Explain what the term 'gas exchange' means.
  b Explain why the amount of nitrogen we breathe in is the same as the amount we breathe out. (*Hint:* nitrogen is an inert gas.)

2 Billy says that the air we breathe out is mainly carbon dioxide. Explain why this is incorrect, and describe to Billy the differences between the air we breathe in and the air we breathe out.

3 Many smokers suffer from emphysema when tar, bacteria and dust build up in the lungs and damage the alveoli. What would the symptoms of emphysema be? Explain your answer.

4 Make a flow chart showing the order of events that occurs when we breathe in and out.

For your notes:

- Oxygen for respiration enters the lungs and passes into the blood through the **alveoli**.

- Carbon dioxide moves from the blood, through the alveoli and into the lungs.

- The movement of gases in the alveoli is called **gas exchange**. The alveoli provide a large surface area for gas exchange.

B4 A transport system

Learn about:
- The circulatory system
- The movement of blood

The transport link

In a city, roads link the places people need to travel to and from. Buses and cars travel along the roads carrying people. Your body, too, has a transport system.

The blood is part of the system which transports reactants and products of respiration to and from all cells. Oxygen and glucose are needed for respiration. The blood collects oxygen from the lungs and glucose from the digestive system. It delivers them to the cells.

The waste products of respiration are carbon dioxide and water. The carbon dioxide is dissolved in the blood and taken away from the cells. The water may be used by the cells, or it may pass into the blood and be taken away.

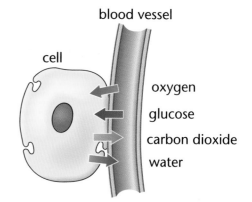

The heart of the matter

The heart, blood vessels and blood are part of an organ system called the **circulatory system**. The heart has valves that let the blood flow in only one direction. A wall in the middle of the heart separates the left and right sides, because each side pumps separately. The left side has a thicker muscular wall to pump the blood all round the body.

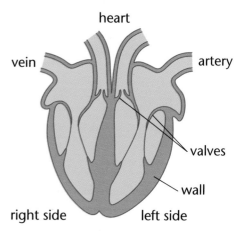

Look at the circulation diagram.

- The left side of the heart pumps blood from the lungs to the body.

- The right side of the heart pumps blood from the body back up to the lungs.

From the heart, the blood goes to the lungs to collect oxygen. Then it goes back to the heart to be pumped around the body to collect glucose from the gut. Blood delivers oxygen and glucose to the cells. The blood collects carbon dioxide and water from the cells. It returns through the heart to deliver them to the lungs.

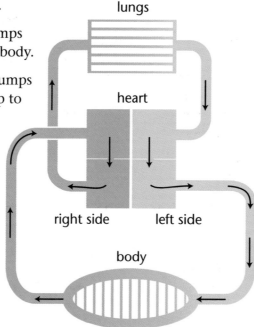

a Why are there valves inside the heart?

Blood vessels

Blood is transported around the body in tubes called blood vessels. There are different types of blood vessel.

- **Arteries** carry blood away from the heart. They have very thick walls made from muscle tissue.

Do you remember?

The muscles in the walls of the heart contract regularly, pumping blood round the body. The blood vessels taking blood from the heart are called **arteries** and those returning it to the heart are called **veins**.

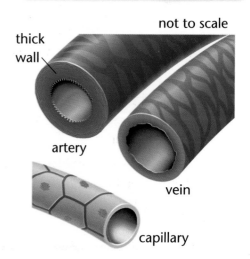

20

- **Veins** carry blood back towards the heart. Veins have valves inside that allow the blood to flow in one direction only. Without valves, gravity would pull all the blood down to your feet!

- **Capillaries** connect arteries and veins. These tiny blood vessels carry blood deep into organs and tissues. The walls of a capillary are made of just a single layer of cells. This allows oxygen and carbon dioxide to pass through easily for respiration.

Ideas have changed

We have known that blood circulates around the body only since the 17th century.

In ancient times, no-one understood how blood flowed around the body. Most theories were based on dissections of dead animals. Aristotle believed that the arteries were full of air because they are usually empty in dead bodies.

In Roman times, a famous doctor called Galen suggested that blood was pumped through the arteries. One idea he had was that blood got into the arteries by leaking through tiny holes in the wall separating the left and right sides of the heart.

In 1543 Versalius showed that there were no such holes by dissecting dead bodies from graveyards.

In 1628 William Harvey experimented on live animals and showed that the blood flowed out of the heart through arteries and came back through veins. He demonstrated on humans that the veins have one-way valves so the blood can never flow backwards. He could only explain this by suggesting that arteries and veins were connected by lots of tiny blood vessels or 'capillaries'. The blood could then go round in a circle.

At the end of the 17th century, after microscopes had been invented, Malpighi saw and described capillaries. Harvey was proved right.

b Why do you think the walls of arteries need to be thick and strong?

Did you know?

It takes blood just 45 seconds to make one trip around the body, and you heart will beat about 25 000 million times in your lifetime!

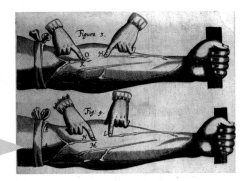

c Draw a table with two columns. Make a list of the different ideas on this page, and the evidence to support each idea.

Questions

1 Explain the function of the circulatory system.

2 Why do you think human circulation is often described as 'double circulation'?

3 Draw a table to show the differences between arteries, veins and capillaries.

4 Describe the journey of blood around the body, starting and finishing at the heart.

5 Draw a timeline to show how ideas and evidence about circulation have changed.

For your notes:

- The blood transports oxygen from the lungs to the cells, and carbon dioxide from the cells to the lungs.

- The heart is a muscular, double pump. One side supplies the lungs; the other side supplies the body organs.

- Ideas about the circulation of blood have changed because of new evidence.

B5 Trouble with yeast

Special brew

Rosie Brown often told this story to her children.

Uncle William was famous in our village for his home-brewed beer. I was introduced to the process at a very early age.

He started by adding a tin of something called hops into a large barrel. Then he added lots of water and sugar. The final ingredient was some rather smelly granules called yeast. He left the mixture in the garage for a few weeks and then bottled it. The yeast made the beer fizz.

One year there was a huge disaster. William put the barrel in the airing cupboard instead of the garage. It fizzed so much that it exploded!

The investigation

Lauren was never quite sure what yeast was. She asked her friend Ryan, and they decided to find an explanation for why yeast makes beer fizz.

Ryan put some water, sugar and yeast in a bottle near the radiator. After a few hours it began to bubble.

Ryan's idea explaining why something happens is called a **hypothesis**.

Lauren was **predicting** what would happen if Ryan's hypothesis was right.

Why does that happen?

I think yeast is a chemical and it is reacting with the glucose to make a gas.

OK, if you are right then the reaction should get faster if we heat up the mixture …

… It's like cooking – chemical reactions happen when food cooks, and food cooks quicker if you heat it up more.

Testing the hypothesis

Ryan and Lauren put the bottle with the yeast mixture on top of the radiator. It bubbled and frothed out of the bottle.

'I was right!' said Ryan. 'Let's see what happens if we boil the mixture in a saucepan first and then put it in the bottle – it should bubble really fast and maybe explode like Great Uncle William's beer!'

They boiled the mixture and carefully placed it back in the bottle on top of the radiator. Much to their surprise, it stopped bubbling completely. They needed to come up with a new hypothesis.

a What was Ryan's hypothesis?

b What was Lauren's prediction?

c What evidence supported Ryan's hypothesis?

d What evidence surprised Lauren and Ryan?

Back in class

Later, Lauren found out that yeast is a tiny fungus made of cells that can only be seen under the microscope. Just like any other living thing, yeast cells release energy from their food by respiration. Another hypothesis sprang to mind. It could be something to do with respiration.

e (i)   What new evidence did Lauren have?
   (ii) What do you think her new hypothesis might have been?
   (iii) How could you test this hypothesis?

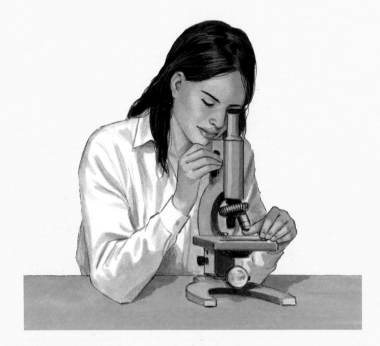

Uncle William used to squeeze the juice of a lemon into his beer barrel. He said it made it fizz better.

f (i)   Can you think of a hypothesis to explain why lemon juice has this effect?
   (ii) Think of a way of testing your hypothesis.
   (iii) Predict what will happen if your hypothesis is correct.

Questions

Discuss these questions with your partner. Write down your answers.

1 What do you think a scientist needs to prove that an explanation is right or wrong?

2 Do you think that one piece of evidence is enough to disprove a theory? Give a reason for your answer.

3 Explain the meanings of these words: hypothesis, prediction and evidence.

4 What do you think Lauren and Ryan should do next?

5 Can you think of any other hypotheses you have tested in science? How did you test them? Were they right?

C1 Going on growing

Unseen microorganisms

Some food goes mouldy if you leave it out in the air. You have probably seen blue-grey mould growing on old bread.

In 1881, Louis Pasteur proved that food decayed because 'germs' we can't see land on it from the air. Germs are tiny living things called **microorganisms**.

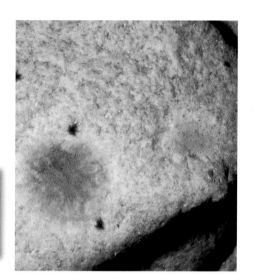

Do you remember?

Microorganisms feed, grow and reproduce like any other organism. Cells reproduce by dividing to make new cells.

A microorganism is a living thing that is so small it can only be seen clearly with a microscope. Microorganisms are sometimes called **microbes**. Many of them are only a fraction of a millimetre long. To describe their size, scientists use a unit called the **micrometre** (μm).

$$1\,\mu m = \tfrac{1}{1000}\,mm = \tfrac{1}{10\,000}\,cm = \tfrac{1}{1\,000\,000}\,m$$

Types of microbe

There are very many types of microbe. There are more microbes on the skin of one person than there are humans on the Earth! There are three main types of microbe:

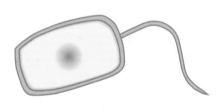

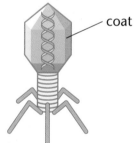
coat

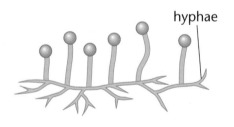
hyphae

Bacteria are very small, usually about 1 μm across. A bacterium is a unicellular organism. Bacterial cells have a cell wall, but do not have a nucleus. They reproduce by cell division.

Viruses are much smaller than bacteria. They are not made of cells.

Fungi (singular **fungus**) are larger than bacteria. Some fungi, such as yeast, are small and round. Others, like mould, are made of long threads called **hyphae**. These threads can only be seen clearly under a microscope.

ⓐ **Some scientists have argued that viruses are not living. Suggest a reason for their argument.**

Useful microbes

The photos opposite show some of the ways that we use microbes.

Fungi are used to make products like those shown below.

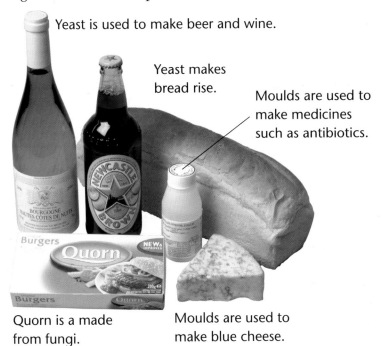

Yeast is used to make beer and wine.

Yeast makes bread rise.

Moulds are used to make medicines such as antibiotics.

Quorn is a made from fungi.

Moulds are used to make blue cheese.

Bacteria are used to make products like those shown below.

Bacteria are used to make yoghurt and cheese.

b Andy's mum told him that all microbes are dangerous. Was she right? Explain your answer.

Growing well

To grow well and reproduce quickly, microbes need to be kept warm and moist and have plenty of food. Microbes respire, but not all microbes need oxygen. For example, yeast can respire without oxygen.

To make wine, yeast is mixed with water and grape juice and left in a warm place. The yeast feeds on the sugar from the grapes, making ethanol. The yeast cells continue to reproduce by dividing into two until they start to compete for food and space. Also, the ethanol builds up and slowly poisons the yeast.

c What do microbes need to grow well?

Waste removal

Bacteria break down the dead bodies of plants and animals, and their waste products. These bacteria are used in sewage farms to break down our sewage. Some bacteria live inside the intestines of animals such as rabbits and help to digest cellulose in their food.

Did you know?

There is a type of bacterium living in your gut called *E. coli*. If the conditions are right, it will divide every 20 minutes!

Questions

1 Draw a diagram of a bacterium, a virus and a fungus. Put them in order of size, smallest first, and label their key features.

2 Make a table listing some uses of fungi and bacteria.

3 Give three differences between viruses and bacteria.

4 Imagine that scientists have discovered a way to kill every single type of microbe on the planet. Write a story about how this would have a serious effect on our lives.

For your notes:

- There are three main groups of **microorganisms**: **bacteria**, **viruses** and **fungi**.

- Bacteria and fungi reproduce by cell division.

- **Microbes** can be very useful to us.

C2 Defence systems

Learn about:
- Microorganisms that cause disease
- How the body fights infection

Disease

The air around you is full of microbes. Your body is covered in them. Many microbes are harmless and most of the time you are healthy. But some microbes can cause **infections** or diseases if they get inside your body. You have probably had a cold or other infections at times. Organisms that cause disease are called **pathogens**.

Do you remember?

Some illnesses, for example chickenpox, food poisoning and colds, are caused by microorganisms. They also cause boils and tooth decay!

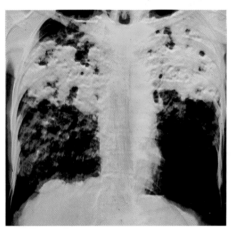

TB is a serious disease caused by bacteria. They have damaged the parts of the lungs shown yellow in this X-ray.

Bacteria attack body cells and release poisonous chemicals (toxins) which kill cells and make you feel ill. Diseases caused by bacteria include tuberculosis (TB), food poisoning, bacterial meningitis and tetanus.

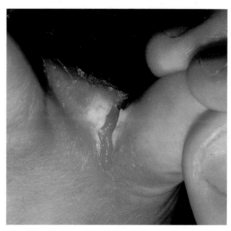

Athlete's foot is a fungal disease.

Fungi often grow on skin and release chemicals that digest skin cells. They can make the skin red and sore. Diseases caused by fungi include athlete's foot, ringworm and farmer's lung.

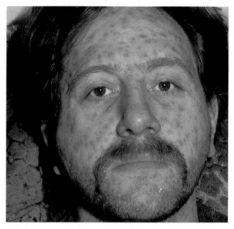

Chickenpox is caused by a virus.

Viruses take over body cells and force the cells to make millions of copies of the virus. The viruses then burst out and invade other nearby cells. They can also release chemicals which make you feel ill. Diseases caused by viruses include colds and 'flu, rabies, chickenpox, German measles, viral meningitis and AIDS.

How microbes enter the body

Your body is very good at keeping microbes out and preventing infection. The skin is a good barrier and stops microbes from getting into the blood. Your tears contain a chemical that destroys bacteria. But there are several ways that microbes can get past these defences and enter your body.

- Cuts in the skin allow microbes in.
- The food you eat can contain harmful microbes.
- The water you drink can carry water-borne microbes.
- Air has lots of microbes in it, which you can breathe in.
- **Sexually transmitted diseases**, such as AIDS, can be caught from sexual intercourse without protection.
- Animals can carry diseases and pass them on by biting you.

(a) Make a table to show examples of diseases that the three types of microbe can cause.

b What natural barriers does the body have against infection?

c Weil's disease is caused by bacteria that are sometimes found in canals and rivers. Explain how canoeists might get infected.

Fighting infection

Once microbes get inside the body, there is still another line of defence which can fight them and protect you. This is called the **immune system**. In the blood, there are **white blood cells** to help in the fight against microbes. They are a vital part of the immune system. They work in three different ways.

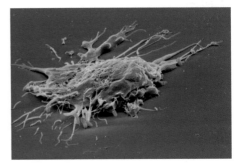

A white blood cell engulfing bacteria (orange rods).

1 Some white blood cells can engulf microbes.

2 White blood cells produce special chemicals called **antibodies** which attach themselves to the outside of the microbes. Antibodies may kill the microbes directly, or they may make them clump together, which makes it easier for white blood cells to engulf them.

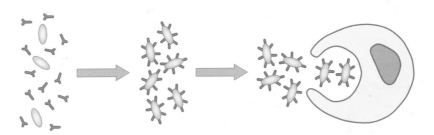

3 White blood cells can destroy the toxins produced by microbes.

Becoming immune

An antibody is only able to recognise and fight one type of microbe. If your immune system has already met a microbe, it can make the antibodies more easily and your body can fight an infection a lot quicker. This makes you **immune** to the disease.

Antibodies can be passed from a mother to her baby across the placenta and also in breast milk. For their first few months after birth, babies are protected from some diseases like measles, because they have antibodies from their mother in their blood.

d Every year the virus that gives you a cold changes slightly. Explain why we catch colds every year.

Questions

1 **a** List four ways disease-causing microbes can enter your body.
  b Describe how bacteria, viruses and fungi can make you ill if they get into your body.

2 Describe how white blood cells protect the body from infection.

3 Why is breast feeding better for babies than bottle feeding?

4 Survey your class to see who has had chickenpox, measles, mumps, colds and 'flu. Find out how many times each person has had each disease. For each disease find out:

  a How many people have had the disease once?
  b How many have had the disease more than once?
  c What does this tell you about this disease?

For your notes:

- Microorganisms that cause **infection** are called **pathogens**.

- The first line of defence in the body is the skin.

- The **immune system** can fight off infection using **white blood cells** and **antibodies**.

C3 Killing bacteria

Learn about:
● Antiseptics
● Antibiotics

Cleaning up

It is important to try to kill bacteria that are around us and on things we touch. In the home, many products contain substances that kill bacteria.

> **Did you know?**
>
> In a fridge or a freezer, the growth of bacteria is slowed down.

Food and drink

We also need to make sure bacteria don't get inside us in our food and drink. Chlorine is added to our drinking water to kill bacteria. Sewage is disposed of safely and treated to prevent it affecting our drinking water.

Cooking food kills bacteria. We use vinegar and salt to preserve food by killing bacteria, and microbes are killed in food before it is canned. Pasteurised milk has been heated to 70 °C for about 15 seconds which kills the bacteria that cause TB.

a Why is it important to dispose of sewage safely?

Disinfectant kills bacteria on floors and in sinks and toilets.

Sterilising solution kills bacteria in babies' bottles.

Antiperspirant kills the bacteria that cause body odour.

Toothpaste kills the bacteria that cause tooth decay.

Antiseptics

We also need to stop bacteria getting inside us through cuts. You can stop a cut becoming infected by putting **antiseptic** cream on it. The cream contains chemicals that kill bacteria.

Tar used to be applied to amputated joints. Then in 1867, the English doctor Joseph Lister thought that it could be microbes that turned wounds bad after operations. He discovered that it was the carbolic acid in the tar that killed bacteria in the air around the wound. So Lister used carbolic acid to kill microbes on wounds and on the instruments he used when operating on people. This made it safer to have an operation because it stopped wounds becoming infected.

b Doctors used to perform operations in their ordinary clothes. Why do you think they wear gowns in modern hospitals?

Antiseptics kill bacteria outside our bodies.

Carbolic acid sprays became widely used during operations after 1867.

Antibiotics

Alexander Fleming made one of the greatest medical breakthroughs by chance in 1928. Fleming was growing bacteria on an **agar plate**. Agar is a jelly used to grow microbes. He noticed that mould growing on the plate stopped a particular type of bacterium from growing. He grew more of the mould and obtained a substance from the broth called **penicillin**. He found it could destroy a number of different bacteria.

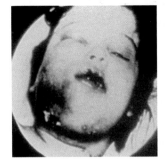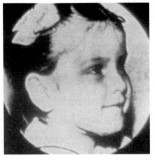

The first child to be treated with penicillin. Four weeks after treatment the infection was gone.

c What was unusual about Fleming's discovery?

Penicillin is an **antibiotic**. Antibiotics are medicines that kill bacteria. None of them will kill viruses. This is why your doctor may give you an antibiotic for a sore throat, but not for chickenpox.

More antibiotics

Since Fleming discovered penicillin, a large number of other antibiotics have been developed. Some of these are shown in this photo.

When your doctor gives you a course of antibiotics, it is important that you take all of them, even if you start to feel better. If you don't take them all, then some of the bacteria will survive and begin to reproduce again. These bacteria may become resistant to the antibiotic.

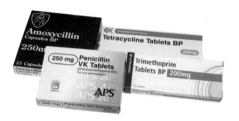

Antibiotics must be prescribed by your doctor.

Each type of antibiotic will kill only certain bacteria so it is important to prescribe the right one. **Broad-spectrum antibiotics** kill a wide range of bacteria. Using a lot of broad-spectrum antibiotics can lead to bacteria becoming resistant to them. **Narrow-spectrum antibiotics** kill a narrower range of bacteria.

d Why should you always finish a course of antibiotics?

Antivirals

Antivirals are drugs used to treat diseases caused by viruses. It is difficult to treat viral diseases because the viruses live inside cells and any drugs that might kill them would also damage the cell. Antivirals can work without killing the virus in three ways by:

- relieving the symptoms of the disease
- stopping the virus reproducing
- strengthening the immune system so that it can fight the virus better.

Questions

1 Describe how you would prevent the growth of harmful bacteria in the home.

2 Describe the medical advances made by each of these scientists:
 a Joseph Lister
 b Alexander Fleming.

3 Explain why people are not prescribed antibiotics if they have colds or 'flu.

4 Explain why it is important to complete a course of antibiotics even if you start to feel better.

For your notes:

- **Antiseptics** contain substances that kill bacteria.

- **Antibiotics** are medicines that kill bacteria, but they have no effect on viruses.

- **Broad-spectrum antibiotics** kill a wide range of bacteria. **Narrow-spectrum antibiotics** kill fewer types of bacteria.

- **Antivirals** are drugs used to treat diseases caused by viruses.

C4 Fighting infection

A helping hand

Once your immune system has met a microbe and made antibodies against it, you are protected. But there is a way to protect you without having to catch every disease to make your immune system work!

You can be **vaccinated** to make you immune to a disease before you catch it. Dead or inactive microbes can be injected into your body. The injection is called a **vaccination**. The microbes do not make you ill, but your body produces antibodies against them. This is called **active immunity** because you have made the antibodies yourself. These antibodies are then ready in case the active microbe ever infects your body. You have been **immunised**.

You can also be injected with ready-made antibodies. Putting ready-made antibodies into your body like this gives you **passive immunity**. Some vaccines providing passive immunity need a booster every couple of years to keep the level of protection high.

a What is the difference between active immunity and passive immunity?

Lady Mary Wortley Montague

In the eighteenth century people were afraid of catching smallpox. People who caught it were very ill. They had sorés full of pus all over their body and usually died.

Lady Montague.

The first way of immunising people against smallpox was brought to Britain from Turkey in 1721 by Lady Montague. She was the wife of the British Ambassador and had survived smallpox herself. In Turkey they took some of the pus from a smallpox sore containing live, active bacteria and put it into a cut made in the vein of a healthy person. Putting microbes into a person is called **inoculation**. It gave them a small dose of the disease and then they recovered and were immune to the active microbe. But unfortunately sometimes things went wrong and the person died.

Lady Montague had her three-year-old son inoculated in Turkey. When she returned to England, she asked a doctor to inoculate her five-year-old daughter. She persuaded her wealthy friends to do the same.

b Why was the Turkish method of immunising against smallpox risky?

Edward Jenner

In 1796, the English doctor Edward Jenner performed the first vaccination against smallpox. Jenner had noticed that milkmaids did not get smallpox, but got a milder form of the disease called cowpox. Jenner took the pus from cowpox spots on a sterile needle and scratched

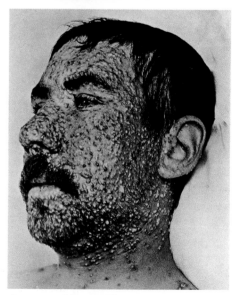

A man with smallpox.

it into the skin of a boy called James Phipps. James developed cowpox, but soon recovered. Later, Jenner inoculated him with smallpox. He did not develop smallpox. He was protected against it.

c **Explain why Jenner's experiment would not be allowed today. Suggest ethical reasons.**

No longer a threat

Jenner gained support for his work from the Royal Family. In 1853, vaccination against smallpox became compulsory for all children in Britain. Smallpox has now been wiped out completely worldwide.

The measles, mumps and rubella vaccination is an example of a continuing programme. This vaccination is called the **MMR vaccination**. It is important to wipe out measles, mumps and rubella because:

- measles can have complications that damage the heart or nervous system
- mumps causes swelling of the salivary glands and can affect the testes in adult males, in some cases making them infertile
- rubella can cause babies to be born blind, deaf or brain damaged if a pregnant woman catches it.

The vaccination appears to have had some side-effects and because of this some parents refuse to have their children vaccinated.

Controlling diseases

Many diseases that are a danger to public health have been controlled by immunising as many people as possible. Polio is now very rare in many countries, thanks to a vaccination programme introduced in the late 1950s.

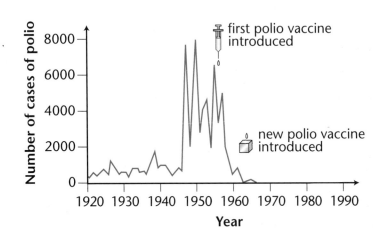

d **Look at the graph. Explain the trend in the number of cases of polio from 1920 to 1970.**

Questions

1 Describe how by Lady Montague helped to prevent people dying from smallpox.

2 Imagine you are James Phipps having a conversation with Edward Jenner. Write down some of the questions you might ask.

3 Why is it particularly important to vaccinate girls against rubella?

4 Some parents decide not to have their children immunised because they are worried about side-effects. What problems might result from this decision?

For your notes:

- The immune system can be helped by **vaccination**.

- Dead or inactive microbes can be injected into your body and your body will produce antibodies against them.

- These antibodies are then ready in case the active microbe ever infects your body.

C5 The battle goes on

The story of Eyam

The nursery rhyme on the right is about bubonic plague. This was a deadly disease caused by a bacterium that infected rats. The fleas that lived on the rats passed the disease to humans when they bit them. It then spread from person to person through coughs and sneezes. People believed that carrying a posy of flowers close to the nose would keep away the disease.

> *Ring a ring of roses*
> *A pocket full of posies*
> *Atishoo! Atishoo!*
> *We all fall down!*

In 1665 the Great Plague spread through London killing thousands of people. When a disease spreads quickly like this there is an **epidemic**. One day, a box of laundry was brought to the village of Eyam in Derbyshire by a traveller from London. The laundry had fleas in it and the plague broke out in the village. More than three-quarters of the people of Eyam died.

The plague would have spread beyond the village, but a brave vicar called William Mompesson persuaded the villagers to stay in their village to contain the disease. Their food and supplies were delivered to the edge of the village. They used to drop money into the well for the delivery people to avoid spreading the infection on the coins.

The village of Eyam today.

ⓐ **What stopped the plague spreading out of Eyam?**

Recent outbreaks

The last plague epidemic happened in the United States in 1924. Since then there have been just a few cases a year in areas with poor housing and rats. Today, people with the disease are isolated in hospital and given antibiotics.

Dr Snow's discovery

If you want to control a disease it is important to find out how it is spread.

In 1848, Dr John Snow worked in London. There was a bad outbreak of cholera in the area. Cholera is caused by a bacterium. At that time, the people got their drinking water from pumps in the street. On a map, Dr Snow showed the pumps and where the victims lived. He found that many of the victims lived near one pump. He suggested that the water from the pump was giving cholera to the people. He closed the pump and there were no new cases.

ⓑ **Look at the map on the right. Which pump do you think Dr Snow closed?**

Cholera is still a problem in areas where people get their water from shared wells or where sewage might get into the drinking water. It is a particular risk after natural disasters such as floods. The spread of cholera in these areas is controlled by:

- adding chlorine to wells
- boiling water
- immunisation
- using antibiotics.

Trouble with mosquitoes

Yellow fever is serious illness caused by a virus found in tropical areas where mosquitoes are active all year round. In 1881, Carlos Finlay suggested that mosquitoes carry the disease from person to person. He was proved right in 1927 when scientists inoculated a rhesus monkey with the blood of a yellow fever patient. They found yellow fever in the monkey.

The disease is controlled with a vaccination, and by spraying the mosquito breeding sites with insecticides.

C (i) **Why do you think Carlos Finlay suggested that mosquitoes carry yellow fever?**
(ii) **Describe two ways of controlling the spread of yellow fever.**

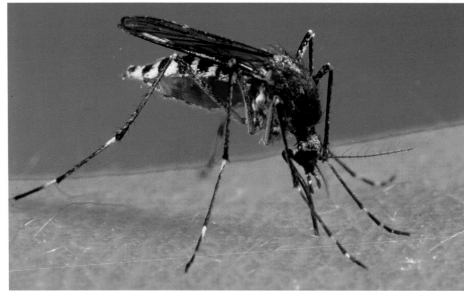

A mosquito bite can inject disease into the bloodsteam.

Questions

1 Explain how bubonic plague is spread.

2 Imagine you are living in Eyam in 1665. Design a poster to warn your fellow residents of the danger they face, with advice about what to do.

3 What modern treatments are used for bubonic plague?

4 Make a time line of all the medical breakthroughs mentioned on these two pages.

5 Why do you think yellow fever is more common in tropical countries?

For your notes:

- Scientists use evidence to tackle the spread of disease in different ways.

- William Mompesson persuaded the Eyam villagers to stay in their village to contain the plague.

- Dr John Snow discovered that cholera was spread by drinking infected water.

- Carlos Finlay suggested that yellow fever was carried from person to person by mosquitoes. The disease was controlled by vaccination and spraying mosquito breeding sites.

C6 Testing medicines

A new medicine

How do scientists design new medicines? They start by looking at what causes disease and how it affects the body. This gives them clues about which chemicals might help to treat the disease.

Before they can be tested on humans, new medicines are first tested on cell cultures. These are collections of cells that have been grown in a laboratory. Successful medicines are then tested on animals to make sure that they are safe to use before trying them on humans.

Clinical trials

Healthy volunteers are given doses of the new medicine. Scientists can then study how the chemical works in the body to check that it is safe to use. After this, the Medicines Control Agency is asked for permission to run trials on patients. These trials show how well the medicine really works. Clinical trials take three to five years.

Correlation

The scientists need to see if the medicine is going to make patients get better. They want to see if there is a link, or a **correlation**, between the patient taking the medicine and getting better.

Scientists usually divide the patients into two groups. One group is given the new medicine, and the other group is given a **placebo**. A placebo acts as a **control** – it does not contain any medicine at all. Neither the patients nor the doctors giving the medicines know which is the real medicine and which is the placebo. This is called a **double-blind trial**.

There are three types of correlation:

- If the medicine works then there is a **positive correlation**.
- If the medicine does not work then there is a **negative correlation**.
- If the numbers of people getting better and not getting better are the same then there is **no correlation**.

a Explain why it is necessary to give some patients a placebo.

b Why do you think a double-blind trial is important?

A new medicine

Dr Franklin was testing a new antiviral medicine to treat a type of influenza. She tested 40 patients. Half were given medicine B182 and the other half were given a placebo. Out of the 20 patients given medicine B182, 15 showed a positive effect (an improvement) and 5 showed no effect. Out of the 20 patients given the placebo, 1 patient showed a positive effect and 19 showed no effect.

To see if there is a correlation between taking medicine B182 and an improvement in the influenza, we use a table like the one on the right.

The 15 people in box **A** proved that the medicine worked because most of the people treated with the medicine got better. The 19 people in box D also proved that the new medicine worked because all but one of the people who were not treated with the medicine did not get better.

	Treated	Not treated
positive effect	A 15	B 1
no effect	C 5	D 19

To find out what the correlation is in the results, we calculate the ratio between A + D and C + B:

A + D : C + B = (15 + 19) : (5 + 1)
      =   34 : 6

If the numbers in the yellow boxes add up to more than the numbers in the brown boxes then there is a positive correlation.

If the numbers in the yellow boxes add up to less than the numbers in the brown boxes there is a negative correlation.

If the numbers are the same then there is no correlation.

c What type of correlation is there for medicine B182?

d What does this test tell you about medicine B182?

e What do you think the reason might be for one person having a positive effect even though they did not receive the medicine?

Further tests for reliability

Dr Franklin then set up another experiment for medicine B182 in another part of the country. She did this to check how reliable her results were. She gave 20 people medicine B182 and another 20 people took a placebo.

Here are the results.

Out of the 20 people using medicine B182, 2 showed signs of improvement and 18 did not.

Out of the 20 people not using the medicine, 18 showed signs of improvement and 2 did not.

f Put these results in a table like the one above.

g Calculate the ratio A + D : B + C.

h What type of correlation does this show?

i What does this test tell you about medicine B182?

Questions

1 Identify the input and outcome variables in this investigation.

2 Why do you think Dr Franklin repeated her experiment in another part of the country?

3 Explain two ways in which Dr Franklin's first experiment shows that the medicine works.

4 Explain two ways in which Dr Franklin's second experiment shows that the medicine does not work.

5 Do you think that the sample size in the trial was big enough? Give your reasons.

6 What do you think Dr Franklin's team should do next?

7 Are there any other factors that scientists should take into account before they give new medicines to people?

D1 Plant groups

Sorting them out

There are millions of species of plants. They make up a large part of the environment providing food and shelter for animals. Scientists classify the millions of plant species into groups to make them easier to study.

Do you remember?

Living things are sorted into groups with similar features. This is called classifying. Animals can be classified as vertebrates or invertebrates.

One way to start to classify plants is by looking at how they **reproduce**. To reproduce, two groups of plants make **seeds** and two groups make **spores**.

Another way is by looking at the leaves. Some plants have leaves with a waterproof waxy layer called a **cuticle** on their top surface to prevent them losing water. We can also look at whether the plants have **veins** to transport water. We call plants that have veins **vascular** plants.

Flowering plants

The biggest group of plants is the **flowering plants**. These include grasses and many trees. Flowering plants reproduce by making seeds inside flowers. Seeds contain food for the new plant, and they can survive in dry or difficult conditions.

Grasses are flowering plants.

There are about 30 000 species of flowering plant on the Earth. Some have large and brightly coloured flowers which are usually pollinated by insects. Others have small, dull flowers which are difficult to spot. These are pollinated by the wind.

Silver birch trees are flowering plants.

Flowering plants have leaves with cuticles and a good water transport system. They are vascular plants. Efficient systems for reproduction and water transport mean flowering plants can grow in a wide variety of habitats. They can grow in dry places where many other plants would die.

a Describe how the flowering plants are adapted to survive in a wide variety of habitats.

Conifers

Conifers are trees with a large trunk and large roots. Conifers make seeds inside **cones**. Their thin needle-like leaves have a small surface area and a cuticle, both of which prevent them losing water. Like the flowering plants, they are vascular plants. Conifers can live in very cold, frozen climates.

Conifers.

Ferns

Ferns reproduce by making small light spores, shown in the photo on the right, which grow into a new fern when they land in a damp place. Ferns need moisture to reproduce.

Ferns have tough leaves called **fronds**. The fronds are large and can trap sunlight well, but they still dry out quite easily in spite of having cuticles. Ferns are also vascular plants. They grow well in damp, cool, shaded woodland habitats.

Ferns are one of the oldest types of plant on Earth. They were around before the dinosaurs, and we often find fossilised ferns.

Mosses

Like ferns, **mosses** reproduce by making small light spores. Look at the photos on the left. Moss needs damp conditions for fertilisation to produce the spores.

Mosses are small plants that look like a springy cushion. They have very small, simple leaves, with no cuticles. They are not vascular so they dry up easily. This means they have to live in wet places. Mosses grow well in boggy habitats where the soil is peat – it is wet and acidic with a layer of partly rotted dead plants.

b Which of the four main types of plant would you expect to find in:
(i) a wet peat bog?
(ii) a shaded area of the garden?
(iii) the slopes of a snow-covered mountain?
(iv) a sunny meadow?

Questions

1 Why would you be more likely to find mosses growing under an oak tree near a duck pond than in the middle of a field?

2 **a** Describe how ferns and mosses reproduce.
  b Explain how the reproduction of ferns and mosses is different from the reproduction of flowering plants and conifers.

3 **a** The pollen from conifers always travels on the wind. How does this make conifers different from flowering plants?
  b Conifers produce large amounts of pollen. Why do you think this is?

4 Which type of forest would you expect to find in the Alps in Switzerland? Give a reason for your choice.

5 Write a key to help scientists put new plants they discover into one of the four main groups. (*Hint:* you could start with how they reproduce or with vascular and non-vascular plants.)

For your notes:

- Plants are classified into four groups by looking at how they **reproduce**, whether the leaves have **cuticles** and whether they are **vascular**.

- **Flowering plants** reproduce from seeds, have cuticles and are vascular.

- **Conifers** reproduce from seeds in **cones**, have thin, needle-like leaves with cuticles and are vascular.

- **Ferns** reproduce from **spores**. They have leaves called **fronds** with cuticles and are vascular.

- **Mosses** reproduce from spores. They have very small leaves with no cuticles and are not vascular.

Ecosystems

A wood has lots of shade. The air is often cool and moist. Trees provide good nesting sites for birds and homes for climbing animals. There are lots of nuts and berries for them to eat in the autumn. The wood provides everything that the organisms in it need for their life processes.

> **Do you remember?**
>
> The place where a plant or an animal lives is called its **habitat**. The surroundings in a habitat are called the **environment**.

An **ecosystem** is a habitat along with all the living things in it, as well as its soil, air and climate.

Different habitats

A habitat might be on land or in water. Examples of land habitats include fields, forests, soil and deserts. Water habitats include ponds, rivers or lakes and even salty water like the sea. Conditions in different habitats vary.

a **Make a list of the different habitats you might find in your school grounds.**

Plants and animals have **adaptations** that help them to survive in their habitat, so we find different plants and animals in different habitats.

Pond life

Becky and Parveen investigated a pond habitat. They wanted to find out whether there was a relationship between oxygen levels and the types of organism found there. Look at the photos on the right. They used sensors connected to a **datalogger** to measure the amount of oxygen in the school pond.

They also used a technique called **pond dipping** to find out what organisms could live in different parts of the pond.

They took sample of water from:

- the surface
- the shallow water at the side
- the deep water
- near the muddy bottom.

They found there was most oxygen in the surface water and in the shallow water at the side. The water near the muddy bottom of the pond had the least oxygen.

On the surface the organism they found most of was the pond skater. The only organisms they found at the bottom of the pond were bloodworms.

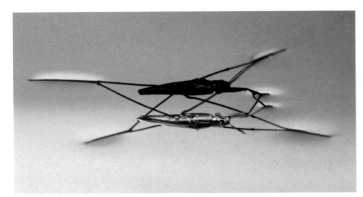

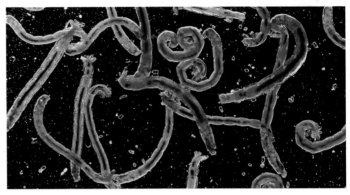

Look at the photo above of the pond skater. It is well adapted to living on the surface of the pond. It is an insect and needs a good supply of oxygen. Its long legs spread its weight out to stop it sinking and its legs are water repellent!

b **What other environmental conditions could Becky and Parveen have measured?**

Sea shells

Gavin and Ian investigated a seashore habitat. They wanted to find out how the animals living on the rocky shore were adapted to survive the force of the waves when the tide comes in.

Most of the animals they found had had thick shells to protect them and some had ways of attaching themselves firmly to the rocks.

Look at the photo above of bloodworms taken under the microscope. They have a red substance called haemoglobin in their blood. This joins with oxygen, so a bloodworm can survive at the bottom of the pond where there is very little oxygen.

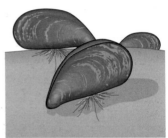

Barnacles have hard chalky shells and cement themselves to rocks.

Mussels make a mass of threads to hold themselves in position.

c **What question did Gavin and Ian choose to investigate?**

Questions

1 The photo on the right shows a very different ecosystem – a city. Describe what this ecosystem is like. Think about:

- the air
- the temperature
- the food available for animals
- the space for animals to live.

2 **a** Explain why a frog and a woodpecker could not swap habitats.
b Under the ground is a perfect habitat for earthworms. Worms have a pointed head to tunnel their way through the soil. It is dark in the soil, so an earthworm does not need to be able to see, but it can sense vibrations and can tell when a bird is walking on the ground above it.

What is the earthworm's best adaptation for avoiding predators?

3 This notice was in the library:

"Enjoy the countryside with the Warwickshire Wildlife Trust. Volunteer helpers are needed to replant woodland areas at several sites near the M42 motorway."

How do volunteers with the Wildlife Trust protect the environment?

For your notes:

- A **habitat** on land or in water has environmental conditions that can be measured.

- Plants and animals have features called **adaptations** that help them to survive.

- An **ecosystem** is habitat together with its living things, the soil, air and climate.

D3 Home alone?

Learn about:
- Pyramids of numbers
- Decomposers

Beachcombing

If you are walking along the beach you might be tempted to see what organisms you can find in a rock pool. You will see different shellfish feeding on the seaweed. If you are patient, you might see the second link in a food chain – a crab ready to eat a mussel.

A food chain shows the feeding relationships between the organisms in an ecosystem. Most food chains start with a plant. The plant makes its own food from carbon dioxide, water and light energy from the Sun. The energy is transferred along the food chain.

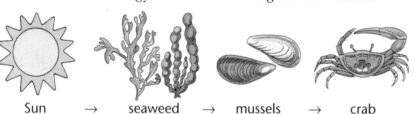

Sun → seaweed → mussels → crab

Pyramids of numbers

Lots of seaweed will feed one mussel. Several mussels will feed just one crab.

If you count the number of crabs, mussels and seaweed plants, you can draw a scale diagram. You can represent the numbers by a bar, so a bar 1 cm long could represent 100 organisms and a bar 2 cm long could represent 200 organisms. We call this a **pyramid of numbers**.

Why are there fewer and fewer consumers as you go up this food chain? This is because energy is lost from the food chain to the environment at each feeding stage. Energy is lost because all the energy transferred to a consumer is not used for growth. When a mussel eats a seaweed, all the energy in the seaweed does not end up in the mussel. Some of it is used for respiration or lost through excretion. Only a small amount is passed on to the crab.

Do you remember?

The arrows in a **food chain** show what eats what, and the direction of the energy transfer.

ⓐ Which organism in this food chain is:
(i) a producer?
(ii) a herbivore?
(iii) a carnivore?

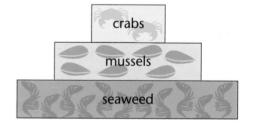

Always that simple?

The diagram for a food chain does not always look pyramid shaped. For example, a single oak tree is very large. It might have more than 10 000 caterpillars on it. A family of bluetits might prey on these, and one owl might eat all the bluetits. The diagram for this food chain shown on the left is not a pyramid shape.

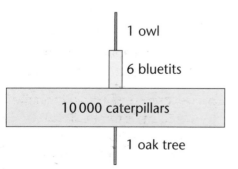

1 owl

6 bluetits

10 000 caterpillars

1 oak tree

ⓑ Why do we get an odd shape when we try to draw a pyramid of numbers starting with an oak tree?

40

Food webs

Lots of food chains in the sea link together to make a **food web**. In a food web, there are several food chains that share the same species. Microscopic animals called animal plankton eat microscopic plants called plant plankton. More than one type of fish eats the animal plankton. In the food web below, both herrings and pilchards eat animal plankton.

c How many different food chains can you find in this food web? Write them out.

Crabs are predators. Mussels are prey for crabs. Humans also like to eat mussels. In some places people catch large numbers of mussels on the beach to sell in fish markets.

d Describe two effects of catching mussels like this on the food web.

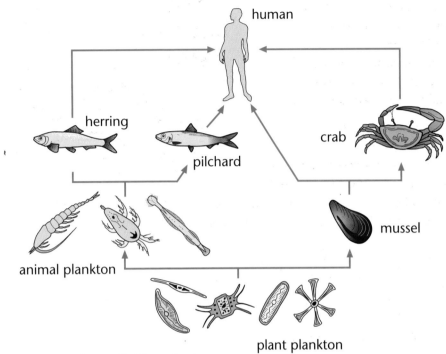

Recycling nutrients

Animals get nutrients by eating plants and animals. Plants need nutrients too. They get minerals from soil or water.

Bacteria and fungi live in rock pools, in rivers and in soil. They feed on the waste products and dead bodies of plants and animals, and break them down. They are called **decomposers**. The decomposers make simpler substances which are returned to the soil or water.

These substances are useful nutrients that plants need to grow. Plants use the decayed waste products and bodies of other plants and animals, and so the cycle begins again.

Questions

1 a Draw a food chain with three links, including humans as one of the links.
   b Make a food web by adding one producer and one herbivore to your food chain.

2 Explain why food chains start with a green plant.

3 A large number of deer once lived on the Kaibab Plateau north of the Grand Canyon in Arizona, USA. In 1907 their predators, wolves, coyotes and pumas, were killed to protect the deer. What do you think happened after this?

4 Decomposers are not always shown on a food web. Explain what they do and why they are important.

For your notes:

● If we count the number of organisms at each level of a **food chain**, we can draw a **pyramid of numbers**.

● In some food chains, a pyramid of numbers is not pyramid shaped, for example if the producer is a large tree.

● **Decomposers** break down the waste and dead bodies of other organisms into useful nutrients that plants need.

D4 Populations

On Onkar

In any habitat, we find lots of species. The number of organisms of a particular species living in a habitat is called the **population**. Read about a population of gimbuls on Onkar.

Conditions on Onkar

The moon Onkar orbits the outer planet of a distant galaxy. Conditions on Onkar are very similar to those on Earth. The intelligent life form is the luhans. They resemble humans but their skins are highly sensitive to ultraviolet light. They live in underground cities away from natural light.

On the ground above the cities, the luhans hunt small mammals called gimbuls to eat. The gimbuls feed on grass and the seeds of the red zetta plant in the early hours of the morning before dawn breaks. They drink water from puddles. Most of the ground is covered with thorny hintel bushes. The thorns protect the bushes from being eaten by the gimbuls.

The gimbuls have adaptations that help them to avoid being eaten by their daytime predators, the wooks. They have large yellow eyes at the sides of their heads for good all-round vision. Their fur has green and red patches that camouflage them against the vegetation. They have large, jagged ears. This means they can see and hear the giant wook birds approaching.

The wooks are also adapted for catching their prey. The aggressive wooks have eyes that point forward for targeting their prey as they get ready to pierce them with their pointed beaks and tear them apart with their sharp claws.

A pair of gimbuls nested in a disused building. There was plenty of dry vegetation among the ruins. The gimbuls ate well and reproduced. They were well hidden from the wooks. The gimbuls had everything they needed. The number of gimbuls in the building grew to a population of 102 after 35 weeks!

Soon things began to go wrong for the gimbuls in the disused building. Death and disease became widespread.

Competition

If the population grows a lot, there is **competition** between the organisms in a habitat. They compete for the resources they need. For the gimbuls:

● the food was running out

● there wasn't enough clean water

● the building was overcrowded and very dirty, so diseases were being passed on.

Wooks on the look-out

A wook had started to notice the gimbuls running in and out of the disused building. He was hungry and ready to attack the gimbuls at dawn. Wooks had always eaten a few gimbuls. Now there were so many gimbuls they became the main food for the wook. Only the gimbuls that were the strongest and fastest escaped the wooks.

a
(i) What do you think happened to the gimbul population in the area?
(ii) Why do you think only the strongest survived the predators?
(iii) What do you think happened to the wook population now that the ailing gimbuls were their main food source?

Predation

A prey animal that is hunted by another animal for food is the target of **predation**. Only the strongest and fittest prey animals survive predation.

A change in population

Look at the table showing how the gimbul population in the building changed over 40 weeks.

Time in weeks	0	5	10	15	20	25	30	35	40
Number of gimbuls	2	8	19	34	65	93	99	102	102

All interdependent

The size of any population depends on competition between the members of that species for food, water and space. It also depends on how many of them are eaten by predators or killed by disease. The different species in a habitat are all **interdependent**. If the gimbul population decreases, so will the wook population.

b Draw a graph of this data.

c Label your graph to show:
- when the population growth was fast
- when the population growth began to slow down
- when the number of gimbuls being born was equal to the number of gimbuls dying.

Questions

1 Explain three ways in which competition for resources can affect the size of a population.

2 On Onkar:
   a which animals are prey?
   b which animals are predators?

3 How do the gimbuls feed?

4 Why are the gimbuls only hunted by the luhans at night?

5 What factors do you think caused the gimbul population growth to slow down?

6 Continue the story of Onkar to describe what happens to the luhans when the population of gimbuls is affected by competition and predation.

7 Draw a food web for Onkar.

For your notes:

- A **population** is the number of individuals of a species living in a habitat.

- **Competition** for resources such as food, water and space, **predation** and disease all affect the size of a population.

D5 Special daisies

Not many left?

Students Joan and Lydia think they have found a rare variety of daisy growing among the ordinary daisies in a field that is to be dug up. The daisy has a bright pink tinge to its petals and grows well in moist conditions. Look at the map of the field.

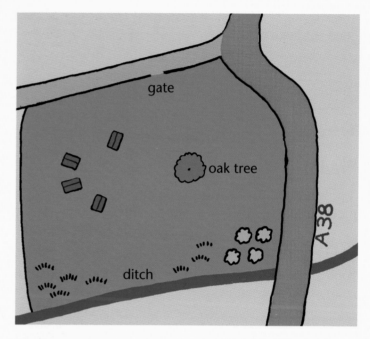

On the map, 1 mm represents 1 m of the field. There are 1000 mm in a metre, so 1 mm on the map means 1000 mm in the field. We call this a **ratio** of 1:1000. Any measurements on the map need to be multiplied by 1000 to give the real distance in the field.

a Measure the distance from the gate to the middle of the largest oak tree on the map.

b Work out the real distance in metres from the gate to the oak tree.

The Northfield Conservation Group wants to find out how many of the daisy plants are in the field. The evidence from their work might save the area from the developers.

Sampling

Joan thought they could count all the rare daisy plants in the field. Lydia suggested that instead of this they should look at a few small areas of the field as samples. 'We could find an area with lots of the daisies, and do our samples there,' she said. 'But that wouldn't be fair,' said Joan. 'It won't tell us the total number in the field.'

They decided that the fairest way of finding out how many of the daisies there are in the field without counting them all was to do **random samples**. This means doing samples in different places without choosing the places deliberately.

c Why do you think they decided not to count every single rare daisy plant in the field?

Joan and Lydia used a **quadrat** to do their sampling. Look at the above photo. This is a wooden frame measuring 1 m on all four sides. This means it has an area of 1 m² (one square metre).

Joan threw a quadrat over her shoulder without looking where it would land. Then she looked at the ground inside the quadrat and counted the number of the daisy plants inside. She threw the quadrat and counted the daisy plants 10 times.

d How did Joan make sure her quadrats were placed randomly?

e How big an area of field did Joan study? (How big was her sample?)

Joan's results

Joan and Lydia found 9 rare daisy plants in Joan's sample.

The whole sample had an area of $10\,m^2$. The field covers $4000\,m^2$ in total. So the ratio of the area they sampled to the whole field is 1:400.

The ratio of the number of plants in the sample to the number of plants in the field is the same, 1:400. So to find the number of plants in the whole field, they multiplied the number in the sample by 400.

f How many plants did they think might be in the field?

g Will this be the exact number of plants in the field?

A bigger sample

Lydia was still worried that most of the quadrats might have landed in parts of the field where there were not many of the daisy plants. This would mean their number for the total rare daisy plants in the field would be lower than the real number. She insisted that they do some more throws to get a more accurate result. So they did 20 new throws, 10 each. Here are the results:

Throw number	Number of rare daisy plants
1	0
2	0
3	0
4	3
5	0
6	0
7	0
8	0
9	0
10	6

Lydia		Joan	
Throw number	Number of rare daisy plants	Throw number	Number of rare daisy plants
1	0	1	0
2	0	2	0
3	3	3	3
4	0	4	0
5	0	5	0
6	0	6	0
7	0	7	4
8	5	8	0
9	1	9	0
10	0	10	0

h How many rare daisy plants did Joan and Lydia find in these 20 new throws?

i What is the ratio of the area of the new sample to the area of the whole field?

j How many plants might there be in the whole field based on the new sample?

Questions

1 a How accurate do you think Joan and Lydia's experiments were?

b How could they have made the experiments more reliable? Think about where the daisy prefers to grow.

2 Do you think the special daisy plants are rare? What other information would you need to help you decide?

E1 It's elementary

Too many

There are too many different materials to count. Just among the solids there are thousands of different plastics, woods, metals, gems, fibres, fabrics and rocks. That doesn't include the liquids and the gases!

How do you make sense of his huge amount of information? It is not easy. It has taken people thousands of years, and we have only made real progress during the last few centuries. This is because most materials are **mixtures** of substances. Some are difficult to separate into **pure** substances.

Gold

People have always been excited by gold. Pure gold is a beautiful, buttery yellow. It never goes black or tarnishes like some metals do. Over the centuries, having gold jewellery meant you were rich and important. The more gold coins you had, the richer you were. Even today, gold still has value all over the world.

Alchemy

Early scientists, called **alchemists**, were always trying to turn other substances into gold. They thought this was the key to making you rich beyond your wildest dreams.

Jabir ibn-Hayyan was a famous Arab alchemist who lived between 760 and 815 AD. He believed that gold could be made from mixing mercury and sulphur. He thought that you mixed mercury and sulphur in different amounts to make different metals. Get the amounts right and you would make gold.

Unfortunately, this isn't true. You can get pure gold from a mixture in rocks, but it has to be there already, mixed up with the other substances. You can even make gold in a chemical reaction, but there has to be some gold particles already there, somewhere.

Elements

Gold is an **element**. If you have a lump of pure gold you cannot break it down into anything simpler. Filtering and distilling have no effect. Even chemical reactions cannot break down gold into anything else. Gold is gold. An element is a substance that cannot be broken down into anything simpler.

Scientists have identified many elements. Mercury and sulphur are elements. So are silver, oxygen, helium and hydrogen.

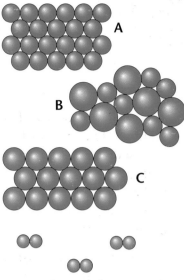

What makes an element?

An element cannot be broken down because it contains only one type of atom. An atom is the simplest type of particle.

a The diagram shows different substances. Two are elements and one is a mixture. Which of A, B or C are the elements?

All materials are made up of atoms. Sometimes these atoms go around on their own, like in helium gas. Sometimes a material is made up of a group of atoms, like in oxygen gas. Oxygen atoms join together in pairs. These groups of atoms are called **molecules**.

The word **particle** can be used for an atom or a molecule, like the word 'child' can be used for a boy or a girl.

helium atoms

oxygen atoms

Elements around us

We find very few pure elements in our surroundings. Gold is found as small lumps of pure gold. Sulphur is found near volcanoes.

Other elements have to be separated from mixtures, or made from other substances by chemical reactions.

'Panning' for gold.

The yellow solid is sulphur.

b Name a mixture that contains the element oxygen.

c Suggest two substances that can be reacted together to make the element hydrogen.

Questions

1 List all the elements mentioned on these pages.

2 A–F all show materials that are pure substances.

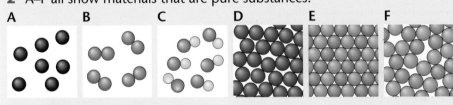
A   B   C   D   E   F

a Which of the substances are:
   (i) gases   **(ii)** liquids   **(iii)** solids?
b Could any of these materials be separated into different substances using filtering, distilling or chromatography?
c Which of the substances could scientists break down using chemical reactions?
d Which of these substances are elements? Explain your answer.

3 Imagine you had a time machine and could go back and talk to Jabir ibn-Hayyan. How would you explain to him that mixing mercury and sulphur could never make gold?

For your notes:

- Most materials are **mixtures**. Mixtures can be separated into **pure** substances.

- Some pure substances are **elements**. Elements cannot be broken down, even with chemical reactions.

- Elements contain only one type of **atom**.

- A **molecule** is a group of atoms joined together. Some elements are made up of molecules.

Learn about:
- Symbols for elements
- The periodic table

Every element has a symbol

People across the world speak different languages. The word for iron is different in each language. But scientists across the world can understand each other because they all use the same **symbol** for iron, Fe. They all know they are talking about the same element.

There are some rules for symbols.

- The first letter is a capital letter.

- If there are second and third letters, they are small letters.

a Imagine that each symbol has only one letter. How many different symbols could there be? Would this be enough?

Some elements have been known from ancient times. The symbol for iron, Fe, comes from the Latin word for iron (ferrum). Latin was the language spoken by the Romans. Gold has the symbol Au, from the Latin word for gold (aurum). Silver has the symbol Ag, from the Latin word for silver (argentum). Tungsten has the symbol W, from the German word for tungsten (wolfram).

b Caesium, calcium, chromium, cobalt, copper, cadmium, cerium, curium and californium are all metallic elements starting with C. Which of the symbols below goes with which element? (*Hint:* the Latin name for copper was cuprum.)

Cm   Cs   Cu   Cd   Cf   Cr   Ca   Co   Ce

Putting the elements in order

Scientists have discovered 113 different elements. That means there are 113 different atoms, because each element contains one type of atom. Learning about all these elements takes a long time, so scientists put them in a special table to make it easier. This table is called the **periodic table**, and most of it is shown on the opposite page. It is used by scientists all over the world.

Look at the vertical columns in the periodic table. They are labelled I, II, III, IV, V, VI, VII and 0. These are called **groups**. The elements in a group are alike. Look at the horizontal rows in the periodic table. They are labelled 1, 2, 3, 4, 5, 6 and 7. These are called **periods**.

Most elements are **metals**. Some elements are not; these are **non-metals**. Some elements have properties of both metals and non-metals, and are more difficult to classify. They are 'between' metals and non-metals.

c Where are the metals in the periodic table? Are they to the left or to the right as you look at the table?

Did you know?

All atoms contain protons. The number of protons is different for each type of atom. Each element has an atomic number, which is the number of protons in that type of atom. It is shown in the periodic table with the symbol.

| O |
| 8 |

Finding a pattern

By 1860 scientists knew about 61 different elements. Discovering an element made you famous. Scientists struggled to find any patterns in the elements.

In 1860 there was the first international meeting for scientists studying chemistry. Some of the scientists started putting the elements in order, using the mass of their atoms. Hydrogen, with the lightest atoms, was first. The race had started to find a useful way of classifying the elements. Different patterns were suggested, but the one suggested by Dmitri Ivanovich Mendeléev was the most useful. He put similar elements in the same group. Mendeléev's pattern became the basis for the periodic table we use today.

					Groups														0	
I	II												III	IV	V	VI	VII		H 1	
																			He 2	1
Li 3	Be 4												B 5	C 6	N 7	O 8	F 9	Ne 10		2
Na 11	Mg 121												Al 13	Si 14	P 15	S 16	Cl 17	Ar 18		3
K 19	Ca 20	Sc 21	Ti 22	V 23	Cr 24	Mn 25	Fe 26	Co 27	Ni 28	Cu 29	Zn 30	Ga 31	Ge 32	As 33	Se 34	Br 35	Kr 36			4
Rb 37	Sr 38	Y 39	Zr 40	Nb 41	Mo 42	Tc 43	Ru 44	Rh 45	Pd 46	Ag 47	Cd 48	In 49	Sn 50	Sb 51	Te 52	I 53	Xe 54			5
Cs 55	Ba 56	La 57	Hf 58	Ta 59	W 60	Re 61	Os 62	Ir 63	Pt 64	Au 65	Hg 66	Tl 67	Pb 68	Bi 69	Po 70	At 71	Rn 72			6
Fr 73	Ra 74	Ac 75	Rf 104	Db 105	Sg 106	Bh 107	Hs 108	Mt 109	Uun 110	Uuu 111	Uub 112		Uuq 114							7

Periods

Key

■ metals   ▨ metal or non-metal

▢ non-metals

Still finding new elements

Teams of scientists are still finding new elements. These elements are very, very rare. That is why it has taken so long to find them. When scientists discover a new element they write about their discovery in a scientific publication. The element is given a temporary name based on its atomic number. The atomic number is converted into words using the key with 'ium' on the end.

So element 111 is called unununium (un-un-un-ium). Its temporary symbol is the first letters of these words, Uuu.

The scientists can only give their element a 'real' name after its existence has been confirmed by a different team of scientists.

zero = nil	one = un	two = bi
three = tri	four = quad	five = pent
six = hex	seven = sept	eight = oct

Questions

1 Lillian thinks that the element F is a metal. Sarah looks at the periodic table and says it is not a metal. Who do you think is right, and why?

2 Joe thinks that all elements should have the first letter of their name as their symbol. Copper should be C and silver should be S. Do you think Joe is right? Give reasons for your answer.

3 a Why isn't there an element 113 or 115?
  b What would be the temporary name and number for element 115?
  c Imagine being the scientist that discovered a new element. What would you choose as its permanent name? Give your reasons.

For your notes:

- There are over one hundred **elements**.
- Each element has a **symbol**.
- We arrange the elements in the **periodic table**.

Properties

When scientists put the elements into the periodic table, they found that the elements in the same group looked and reacted in similar ways. The appearance of elements and the way they react are called their **properties**.

The first thing you notice when you study elements is their appearance and if they are solid, liquid or gas. These are **physical properties**.

Appearance

Kay's class was looking at the appearances of elements. The class agreed that metals can be shiny, but Ross pointed out that you get rusty and dull metals. They decided that if you polish a metal it is shiny.

Some metals.

They also noticed that all the non-metals have very different appearances.

a Look at the photos. What is the difference between the appearance of metals and non-metals?

Some non-metals.

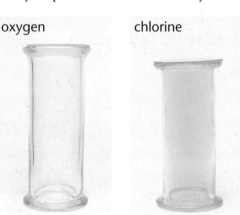

oxygen   chlorine

bromine

sulphur

Solid, liquid or gas?

Most of the elements in the periodic table are solids at room temperature.

b Look at the diagram of the periodic table.
   (i) Are all the metals solids?
   (ii) How many elements are gases?

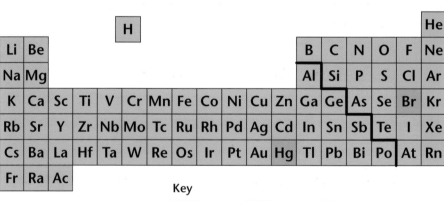

Key

solid   liquid   gas

When you heat a solid metal, it melts and becomes a liquid. Zinc melts at 420°C and gold at 1064°C. These temperatures are their **melting points**. Mercury is a liquid. You would have to cool mercury to make it a solid. Its melting point is −39°C.

Bromine is a non-metal that is liquid at room temperature. If you heat it up it boils and becomes a gas. Bromine boils at 59°C. This temperature is its **boiling point**.

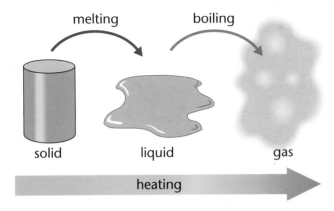

melting   boiling

solid   liquid   gas

heating

c (i) How many elements are liquid at room temperature?
(ii) Name two non-metals which are solid at room temperature.
(iii) Name two non-metals which are gases at room temperature.

Magnetic or not?

Katie says that all metals are **magnetic**, but Ross and Kay disagree. They test lots of elements. They show that only iron, nickel and cobalt are attracted to a magnet. You will learn more about magnetic materials in unit J.

Chemical reactions

The ways that elements change during chemical reactions are called **chemical properties**. Many elements react in a similar way in chemical reactions.

One example of this is when an element reacts with oxygen to make an oxide. Metals and non-metals react in this way. For example, carbon and oxygen react to make carbon dioxide, and magnesium and oxygen react to make magnesium oxide.

d **What will be made when copper reacts with oxygen?**

Do you remember?

All metals conduct electricity. They allow electricity to flow easily.

Do you remember?

Melting and boiling are examples of **physical changes**. They are reversible changes. No new substances are made in a physical change.

Chemical reactions are **chemical changes**. They are irreversible changes. New substances are formed.

Questions

1 Use the information on these pages to decide if these statements are 'true' or 'false'. Give at least one piece of evidence for each answer.

   a All metals are magnetic.   **b** All metals conduct electricity.
   c All metals are always shiny. **d** All metals are solids.

2 In what three ways are most non-metals different from most metals?

3 What is unusual about these elements?

   a mercury   **b** bromine   **c** nickel

4 Many non-metals are gases at room temperature. Compare the boiling points of these non-metals with the boiling points of metals. Explain your answer.

For your notes:

● Elements have different **physical** and **chemical properties**.

● Elements may be solid, liquid or gas.

● Most metals are solid and shiny when polished. A few metals are **magnetic**.

● Non-metals are mostly solid or gas. They have many different appearances.

Metal or non-metal?

Scientists have used the properties of metals to decide that:

● iron, copper, nickel and mercury are metals

● carbon, sulphur, selenium, bromine and chlorine are non-metals.

Joe's class is thinking about how the scientists decided on these two groups. They want to work out the rules for deciding if an element is a metal or a non-metal. They use the fact files on the opposite page.

Lillian's idea

Put all the solids in one group. Metals are solids. All the others will be non-metals.

Joe's idea

Test the elements with a magnet. Those that stick to the magnet are metals. The others are non-metals.

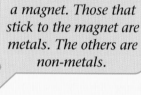

a (i) Which elements will be in Lillian's 'metals' group?
(ii) Which elements will be in Lillian's 'non-metals' group?
(iii) Is Lillian's idea going to work? Explain your answer.

b (i) Which elements will be in Joe's 'metals' group?
(ii) Which elements will be in Joe's 'non-metals' group?
(iii) Is Joe's idea going to work? Explain your answer.

Yasmin's idea

See which elements conduct electricity. Those that do are metals. Those that don't are non-metals.

Tony's idea

The shiny ones are metals. All the others are non-metals.

c (i) Which elements will be in Yasmin's 'metals' group?
(ii) Which elements will be in Yasmin's 'non-metals' group?
(iii) Is Yasmin's idea going to work? Explain your answer.

d (i) Which elements will be in Tony's 'metals' group?
(ii) Which elements will be in Tony's 'non-metals' group?
(iii) Is Tony's idea going to work? Explain your answer.

e Which were the most successful of the four ideas?

f Decide on two questions that will group the elements into metals and non-metals.

Fact files

The metals are in red boxes. The non-metals are in yellow boxes.

Element	Copper
Symbol	Cu
State at 25°C	solid
Colour	pink
Shiny?	✓
Magnetic?	✗
Conduct electricity?	✓

Element	Iron
Symbol	Fe
State at 25°C	solid
Colour	grey
Shiny?	✓
Magnetic?	✓
Conduct electricity?	✓

Element	Nickel
Symbol	Ni
State at 25°C	solid
Colour	grey
Shiny?	✓
Magnetic?	✓
Conduct electricity?	✓

Element	Mercury
Symbol	Hg
State at 25°C	liquid
Colour	silver
Shiny?	✓
Magnetic?	✗
Conduct electricity?	✓

Element	Selenium
Symbol	Se
State at 25°C	solid
Colour	silver
Shiny?	✓
Magnetic?	✗
Conduct electricity?	✗

Element	Carbon
Symbol	C
State at 25°C	solid
Colour	black
Shiny?	✗
Magnetic?	✗
Conduct electricity?	✓

Element	Sulphur
Symbol	S
State at 25°C	solid
Colour	yellow
Shiny?	✗
Magnetic?	✗
Conduct electricity?	✗

Element	Bromine
Symbol	Br
State at 25°C	liquid
Colour	brown
Shiny?	✗
Magnetic?	✗
Conduct electricity?	✗

Element	Chlorine
Symbol	Cl
State at 25°C	gas
Colour	green
Shiny?	✗
Magnetic?	✗
Conduct electricity?	✗

Questions

There are some elements that do not fit neatly into the group 'metals' or the group 'non-metals'.

1 Look at this information about silicon, phosphorus and boron. Which are metals and which are non-metals? Give reasons for your decisions.

2 Look back at the periodic table on page 49. How do the scientists classify silicon, phosphorus and boron? Do you agree with them?

Element	Silicon	Phosphorus	Boron
State at 25°C	solid	solid	solid
Colour	grey	black	grey
Shiny?	✓	✗	✗
Magnetic?	✗	✗	✓
Conduct electricity?	✗	✗	✗

E5 Reacting elements

Burning elements

Kay's class is burning elements. When you burn something it reacts with oxygen in the air. The teacher burns magnesium and it reacts with oxygen. She asks the class to look for evidence that a chemical reaction is happening.

a Do you think a chemical reaction has taken place? Give reasons for your answer.

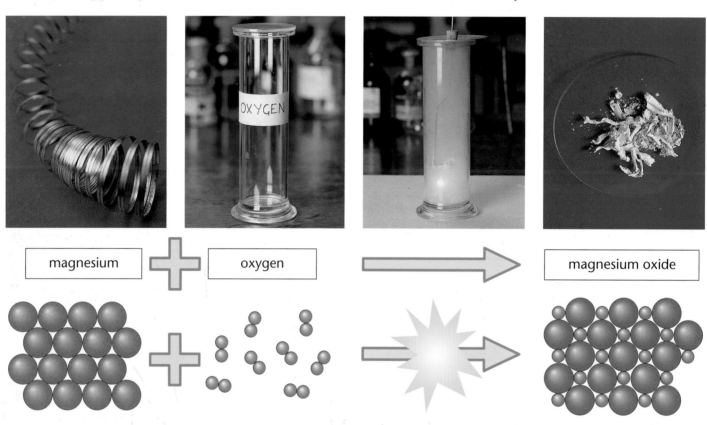

| magnesium | ✚ | oxygen | ➡ | magnesium oxide |

New substances

When magnesium burns in oxygen an oxide is made. It is a new substance called magnesium oxide. Magnesium oxide is not an element. It is made up of more than one type of atom. It is a **compound**. Elements react together to make compounds. Oxides are one type of compound.

b What compound will form when oxygen reacts with:
(i) calcium?   (ii) silver?

Look at the diagram above showing the particles in magnesium oxide. Each magnesium atom is surrounded by oxygen atoms. Each oxygen atom is surrounded by magnesium atoms. This is because the magnesium atoms and the oxygen atoms were joined together during the chemical reaction.

Every compound has a formula

It would take too long to draw particle diagrams every time we want to show the atoms in compounds. We can use the symbols to represent the atoms. We write **MgO** to represent magnesium oxide.

MgO is the **formula** for magnesium oxide. It tells us that magensium oxide is a compound, not an element, and that it contains magnesium atoms and oxygen atoms.

c Which of these substances are compounds?

Na   CaO H_2O CO_2 O_2 Au   Ag

Predicting how elements react

We have already seen that when elements burn in oxygen, they react to make oxides. Elements that react with sulphur all react in a similar way to make sulphides.

Hydrogen reacts with sulphur to make hydrogen sulphide. The hydrogen and sulphur atoms react and join together. Hydrogen sulphide is a compound.

d **Magnesium and sulphur react in a similar way. Write the word equation for the reaction.**

You have seen that oxides all have oxygen in them, and sulphides all have sulphur in them. There are other names with similar patterns to help us know what elements are present in compounds. Some examples are shown in the table.

e **Predict the products of the reactions of magnesium with each of these elements and write the word equation for each reaction.**

(i) **chlorine**   (ii) **bromine**   (iii) **fluorine**.

hydrogen + sulphur   →   hydrogen sulphide

Compound contains	Compound name
oxygen	oxide
sulphur	sulphide
chlorine	chloride
bromine	bromide
fluorine	fluoride

Questions

1 **a** Which two elements would you react together to make:
   (i) copper oxide   (ii) calcium chloride   (iii) iron bromide?
   b Write word equations to show the three reactions in **a**.

2 Predict the products of these reactions and write a word equation for each:
   a zinc and oxygen
   b potassium and chlorine
   c iron and fluorine
   d sodium and sulphur.

3 Name these compounds:
   a FeO   **b** CuF_2 **c** Ag_2S **d** $MgCl_2$ **e** $AlBr_3$

4 Sulphur reacts with oxygen to make two different compounds, sulphur dioxide and sulphur trioxide.
   a Which of these formulae is sulphur trioxide?
   $$SO_2 \qquad SO_3$$
   b Which of the two compounds will be made when there is lots of oxygen present? Give reasons for your answer.

Did you know?

The pictures of atoms on these pages are 25 million times bigger than the real atoms.

For your notes:

- Elements react together to make **compounds**.

- A compound is a substance with more than one type of atom joined together.

- A compound is represented by a **formula** that shows the ratio of the different atoms in the compound.

F1 Compounds all around

Learn about:
- The range of compounds
- The properties of compounds

Elements and compounds

In Unit E Atoms and elements you learnt that an element is a substance that contains only one type of atom and a compound is a substance with more than one type of atom joined together.

Is it water?

Water is a compound. The diagram below shows two molecules. The molecule on the left is a water molecule. The molecule contains two hydrogen atoms and one oxygen atom. The white balls stand for hydrogen and the red ball stands for oxygen.

The molecule on the right in the diagram is a **hydrogen peroxide** molecule. Hydrogen peroxide is also a compound. It also contains hydrogen and oxygen atoms joined together.

a How many different atoms are there in water?

b Why isn't water an element?

c What is the difference between the water molecule and the hydrogen peroxide molecule?

Do you remember?

A molecule is a group of atoms joined together. They can be the same atom, as we saw in an oxygen molecule, or they can be a group of different atoms, like in carbon dioxide.

Did you know?

One extra oxygen atom makes a big difference. Hydrogen peroxide is a dangerous, corrosive liquid. Diluted with water it is used as a disinfectant.

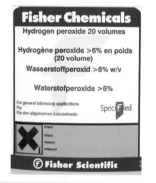

Glucose

Glucose is also a compound. Look at the diagram on the right. It shows a model of one glucose molecule. The formula for glucose is $C_6H_{12}O_6$.

As you can see, the glucose molecule is larger than water. It contains 48 atoms joined together.

Sodium chloride

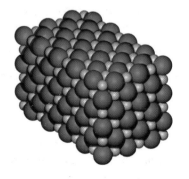

Not all compounds are made of molecules. In some compounds the different atoms are arranged to make crystals rather than molecules. The scientific name for salt is sodium chloride. Each sodium chloride crystal contains millions of atoms. The diagram on the left shows sodium atoms (grey) and chlorine atoms (green) arranged to make a crystal.

The formula for sodium chloride is NaCl. This shows there is one sodium atom for each chlorine atom in sodium chloride.

d How many different types of atoms are there in:
   (i) carbon dioxide?   (ii) glucose?   (iii) sodium chloride?

Hazard to harmless

The compound sodium chloride is very different from the two elements called sodium and chlorine.

Sodium is a shiny, silver metal. It has to be stored under oil because it would lose its shine and turn into a white powder if you left it out in the air.

Chlorine is a poisonous green gas.

Sodium chloride is a solid with colourless crystals.

Compare the properties

This table compares sodium, chlorine and sodium chloride.

The melting point is the temperature at which a solid melts or a liquid freezes. The boiling point is the temperature at which a liquid boils or a gas condenses. Each **pure** substance has a known melting point and boiling point.

You can see that sodium chloride has a much higher melting point than either sodium or chlorine.

Substance	Sodium	Chlorine	Sodium chloride
Element/compound	element	element	compound
Symbol/formula	Na	Cl	NaCl
Type	metal	non-metal	salt
Appearance	shiny, silver solid	green gas	colourless crystals
Melting point in °C	98	–101	801
Boiling point in °C	883	–35	1413
Does it react with hydrochloric acid?	Yes, it makes hydrogen.	no	no

e Compare the boiling points of sodium, chlorine and sodium chloride.

Questions

1 Look at these diagrams. They show models of three different compounds. Give the formula for each. (Black is carbon, red is oxygen, white is hydrogen, blue is nitrogen.)

A   B

C

2 **a** Use the information on these two pages to make a list of the differences between:
   (i) sodium and sodium chloride
   (ii) chlorine and sodium chloride.

   b Write a word equation for the formation of sodium chloride from sodium and chlorine.

   c How do you know that a new substance is made when sodium and chlorine react?

For your notes:

- Some compounds contain molecules.

- Other compounds have different types of atoms fixed together to make crystals.

- Compounds are very different from the elements that made them.

- All **pure** substances have a known melting point and a known boiling point.

In a flash

Kevin's class is looking at reactions involving compounds. His teacher, Mrs McMichaels, shows them a reaction between copper oxide, a compound, and zinc, an element.

The word equation for the reaction is:

zinc + copper oxide → copper + zinc oxide

One compound, copper oxide, has reacted to make another compound, zinc oxide.

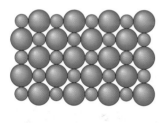

 oxygen copper oxygen zinc

Look at the diagram above showing the particles in copper oxide. Compare it to the diagram above right showing the particles in zinc oxide.

(a) Look at the photo above showing the reaction. How do you know that a chemical reaction has taken place?

(b) What are:
(i) the reactants?
(ii) the products of this reaction?

(c) Where did the zinc atoms come from to make the zinc oxide?

(d) Where did the copper atoms go to when the copper oxide reacted?

All change

Kevin and Bianca then carry out a second reaction. They take some sodium iodide and they dissolve it water. It makes a colourless solution. Then they take some lead nitrate and dissolve that in water. It makes another colourless solution.

Then they mix the two solutions.

The reactants are sodium iodide and lead nitrate. The products are lead iodide and sodium nitrate. The lead iodide is the yellow solid. The lead iodide is a **precipitate**. A precipitate is a solid that is made when two liquids react.

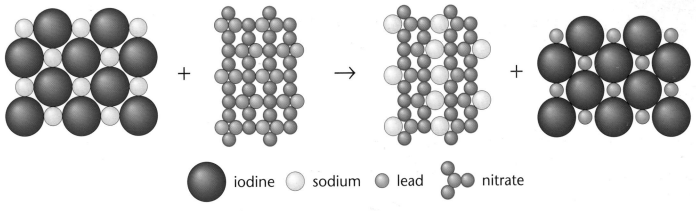

iodine ⚫ sodium ⚪ lead ⚪ nitrate ⚛

e What two observations show that a chemical reaction has taken place?

f Write a word equation for this reaction.

g Use the diagram above to explain what happens when sodium chloride and lead nitrate react.

Bubble

Chemical reactions also happen in cells. Look at the photo on the right. The pea is in a solution of hydrogen peroxide. The cells change the hydrogen peroxide into water and oxygen.

h What evidence, shown in the photo, shows that a chemical reaction is taking place?

The diagram below shows the compounds and elements involved in this reaction.

hydrogen peroxide   →   water   +   oxygen

Questions

1 Which reaction, described on these two pages showed:

  a a colour change?
  b energy given out?
  c bubbles being made?
  d a precipitate?
  e energy being taken in?

2 Look back at the diagram showing how the plant cells change hydrogen peroxide molecules.

  a What is the formula of:
     (i) hydrogen peroxide? **(ii)** water?   **(iii)** glucose?
  b What element is shown in this diagram?
  c How many oxygen atoms are there in:
     (i) the reactant molecules?   **(ii)** the product molecules?
  d Predict, without counting, how the number of hydrogen atoms changes during the reaction.

For your notes:

● Compounds change into other compounds, or make elements, in chemical reactions.

● A **precipitate** is a solid made when two liquids react.

59

Water, water everywhere

Bianca, Ellie and Kevin are investigating water. They collect samples of different types of water. They have been asked to find out if the water is pure.

Do you remember?

Pure means there is only one substance present. Pure water contains water only.

Using the label

They start by investigating the mineral water, because it has a label.

a Look at the label. Name three substances the mineral water contains other than water.

NATURAL MINERAL WATER

1.5 LITRE e

TYPICAL ANALYSIS mg/L:

Ca	35.0	HCO$_3$	136.0	F	<0.1
Mg	8.5	Cl	7.5	Fe	<0.01
Na	6.0	SO$_4$	6.0	Al	<0.01
K	0.6	NO$_3$	<0.1	T.D.S at 100°C	136

CALORIE FREE   pH at source 7.8
LOW MINERAL CONTENT.
SUITABLE FOR A LOW SODIUM DIET.
ONCE OPENED STORE IN REFRIGERATOR
AND USE WITHIN 7 DAYS.

*The mineral water is not pure. It is a **mixture**. Look at all the other stuff it contains.*

What's left?

They measure out 100 cm^3 of seawater and boil it dry in an evaporating dish.

b Predict what will be left when the water is evaporated from the seawater.

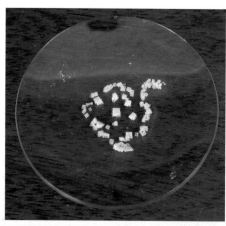

When all the water is evaporated they are left with a white, powdery solid. This is a **mixture** of sodium chloride, salt, and other substances.

Bianca then looks back at the mineral water label. It says 'Dry residue at 180°C 130 mg/litre'.

c If 1000 mg = 1 g and 1000 cm^3 = 1 litre, how much powder would be left in the evaporating dish if Bianca evaporated:
(i) 1000 cm^3 mineral water?
(ii) 100 cm^3 mineral water?

Do you remember?

Filtration, evaporation, distillation and chromatography are all ways of separating.

Distilled

Ellie asks Mrs Cook, the science technician, how the distilled water is made. She explains that it is made from tap water by distillation. The tap water is heated until it boils. The water turns into steam, leaving any other substances behind. The steam is collected and condensed.

d **Do you think the distilled water is pure? Explain your answer.**

Other mixtures

Most of the materials around us are mixtures. Petrol, washing powder, paint and paper are all mixtures. That is why we can have different types of petrol, different washing powders, different coloured paint and different types of paper.

Constant or varied?

If a substance is a compound, you know exactly what is in it. Scientists say that compounds have a **fixed composition**. That means that they are always made up from the same atoms, present in the same ratio.

Pure water contains water molecules only. Each water molecule contains two hydrogen atoms and one oxygen atom. This means that the ratio of hydrogen atoms to oxygen atoms in water is 2:1. This is shown in the formula for water, which is H_2O.

e **The formula for calcium chloride is $CaCl_2$. What types of atoms does calcium chloride contain?**

The mineral water contains lots of different substances. Different types of mineral water contain different substances. The seawater is different again. So are tap water, rainwater and river water.

Mixtures vary, compounds are always the same.

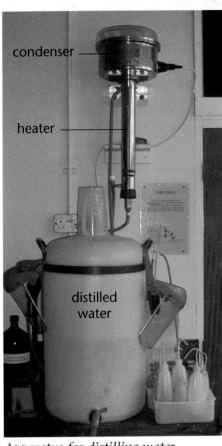

Apparatus for distilling water.

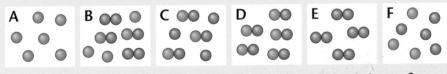

Questions

1 Decide if the following statements are *true* or *false*.

  a Mixtures contain only one substance.
  b An element is a mixture.
  c A mixture of different elements is a compound.
  d Mixtures have a fixed composition.
  e Compounds have a fixed composition.
  f Elements have a fixed composition.

2 List four methods of separating mixtures.

3 Look at this diagram showing the particles in six different gases.

  a Which gases are pure?
  b Which gases are mixtures?
  c Which gases are elements?
  d Which gases are compounds?

For your notes:

- **Mixtures** contain more than one substance.

- Pure substances contain only one element or one compound.

- Mixtures vary. Compounds have a **fixed composition**.

F4 The air around us

What's in air?

Air is a mixture of nitrogen, oxygen and other gases. The particles in air are shown in the diagram on the right and the proportions of the different substances in air are shown in the pie chart.

Air does not always contain exactly the amounts shown in the pie chart. Air near a bonfire will contain more carbon dioxide and some smoke. Air in a meadow will contain the chemicals made by flowers to attract bees.

The air in a crowded lift will contain less oxygen than in an empty lift because the people will be taking oxygen out of the air. Air varies, it does not have a fixed composition. This is because it is a mixture.

ⓐ Which particles in air are:
 (i) molecules?
 (ii) single atoms?

ⓑ Use the pie chart to describe the composition of air.

ⓒ Explain how air can vary from the composition shown in the pie chart.

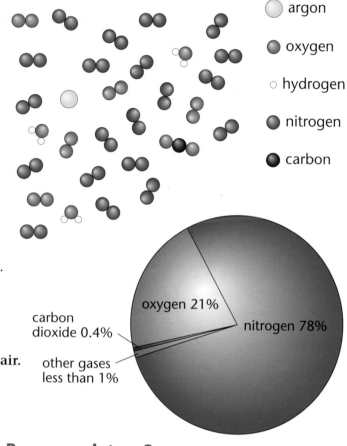

○ argon
● oxygen
○ hydrogen
● nitrogen
● carbon

carbon dioxide 0.4%

oxygen 21%

nitrogen 78%

other gases less than 1%

The air is a mixture. The nitrogen is a pure substance.

Water boils and condenses at 100°C. What temperature does nitrogen boil and condense at?

If the scientist cooled the nitrogen gas, it would condense at −196°C.

Pure or mixture?

Mrs McMichaels sets her class a challenge. She asks them to find out how a scientist could tell the difference between air and nitrogen.

Kevin finds out that pure substances boil at one temperature, the boiling point, and melt at another temperature, the melting point.

Ellie looks up the boiling point of nitrogen. It is −196°C.

They then think what would happen with the air. The air contains mostly nitrogen and oxygen. The boiling point of oxygen is −183°C. They imagine the scientist cooling the air. When you cool a gas it will condense into a liquid. So, when you cool the air the oxygen would condense at −183°C. Then the nitrogen would condense at −196°C.

They go on an 'Ask the scientist' website to check their idea. The scientist, Dr Taylor, sends back the graphs at the top of the next page.

d Explain how cooling the gases can be used to tell air and nitrogen apart.

e Why does Dr Taylor say '*at least* three different substances' rather than 'three different substances'?

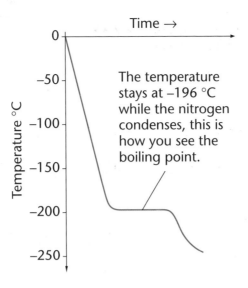

The temperature stays at –196 °C while the nitrogen condenses, this is how you see the boiling point.

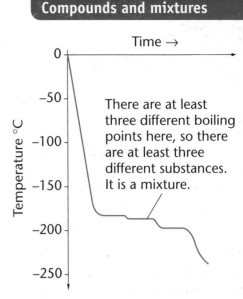

There are at least three different boiling points here, so there are at least three different substances. It is a mixture.

Separating air

In hospitals oxygen is added to the air to make it easier for some patients to breathe. Pure oxygen is made by cooling air to –200°C. This means all the nitrogen and all the oxygen will have condensed. It will be liquid air.

The liquid air is then heated up. The nitrogen boils first. All the nitrogen boils away. The oxygen boils next, making oxygen gas.

This is a type of distillation. It is carried out on a very large scale to make all the oxygen we need in hospitals and in industry.

Nitrogen is also very useful. Liquid nitrogen is used to keep things very, very cold. Cells kept at –200°C stay almost perfect. Embryos kept in liquid nitrogen for years can be defrosted and go on to become healthy babies.

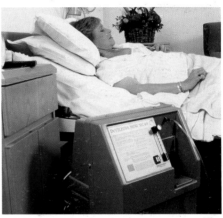

This machine concentrates oxygen from the air to treat the patient.

f During the distillation of air, at what temperature does:
  (i) the nitrogen boil?
  (ii) the oxygen boil?

Questions

1 Lesley and Clare find out about a different way of telling which gas is air and which is nitrogen. They find out about a 'gas chromatography' machine. This separates the gases in the mixture, like chromatography uses paper to separate coloured dyes.

This diagram shows the results of putting two gases through the 'gas chromatography' machine.

a Which gas is nitrogen and which is air?

b Explain how you came to your decision.

A

B

2 Use the information on these two pages to give as full an answer as you can:

  a How do we separate air?   **b** Why do we separate air?

F5 Formulae

Ratios

Scientists use **ratios** to work out the formula for compounds.

Look at the diagram on the right. It shows the compound sodium chloride. There are 25 atoms of sodium and 25 atoms of chlorine. We can write this:

sodium : chlorine
25 : 25
1 : 1

The *ratio* of sodium atoms to chlorine atoms is 1:1. This is why the formula for sodium chloride is NaCl.

Look at the diagram on the right. It shows the compound calcium chloride. There are 12 atoms of calcium and 24 atoms of chlorine.

calcium : chlorine
12 : 24
1 : 2

The ratio of calcium atoms to chlorine atoms is 1:2. This is why the formula for calcium chloride is $CaCl_2$.

Look at the diagram on the right. It shows the compound sodium oxide. There are 20 atoms of sodium and 10 atoms of oxygen.

sodium : oxygen
20 : 10
2 : 1

The ratio of sodium to oxygen atoms is 2:1. This is why the formula for sodium oxide is Na_2O.

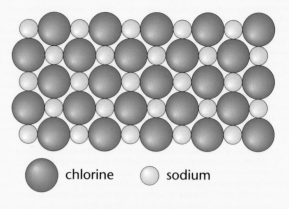

chlorine   sodium

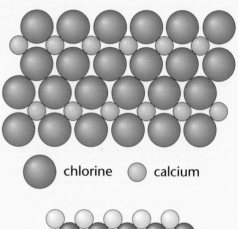

chlorine   calcium

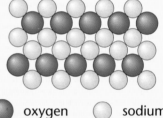

oxygen   sodium

ⓐ **What is the formula for:**

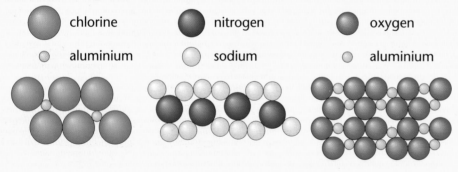

chlorine   nitrogen   oxygen

aluminium   sodium   aluminium

   (i) **aluminium chloride?**   (ii) **sodium nitride?**   (iii) **aluminium oxide?**

ⓑ **(i)** **If some sodium chloride contained 20 million atoms of sodium, how many atoms of chlorine would it contain?**

    (ii) **If some sodium oxide contained 600 billion atoms of sodium, how many atoms of oxygen would it contain?**

Constant compound

Craig's class are doing an experiment. They are burning different masses of magnesium in air. They weigh the amount of magnesium oxide they make. They take great care to make sure that:

● all the magnesium is burned

● none of the magnesium oxide escapes.

They then work out the mass of oxygen that has reacted with the magnesium to make the magnesium oxide. Their results are shown in the table and the graph below.

Mass of magnesium in g	Mass of magnesium oxide in g	Mass of oxygen in g
1.00	1.92	0.92
2.00	3.09	1.09
3.00	3.35	0.35
4.00	6.41	2.41
5.00	8.52	3.52
6.00	9.71	3.71
7.00	11.18	4.18
8.00	13.49	5.49
9.00	14.83	5.83

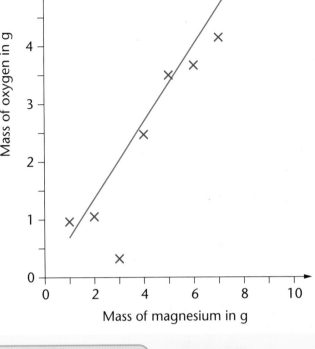

c **Which result does not fit in with the others?**

There is a straight line of best fit. This shows that the ratio of the mass of magnesium to the mass of oxygen stays the same. The **composition** of magnesium oxide is always the same.

d **Using the graph above, how much oxygen reacts with:**
(i) 3 g magnesium?
(ii) 6 g magnesium?
(iii) 9 g magnesium?

e **What is the ratio of the mass of magnesium to the mass of oxygen?**

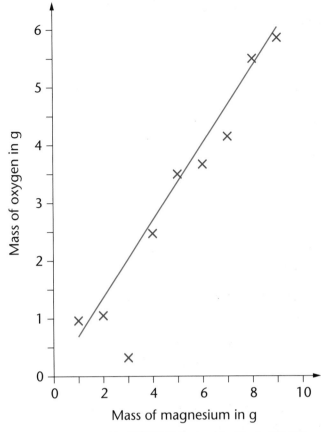

24 units Mg

16 units O

I thought the formula for magnesium oxide was MgO. That's a ratio of 1:1. The ratio of mass of magnesium to mass of oxygen is 3:2.

Well spotted! Magnesium atoms and oxygen atoms do not have the same mass. Magnesium atoms weigh more than oxygen atoms.

Questions

1 Draw diagrams of these solids showing how the atoms may be arranged.

a MgO **b** $MgCl_2$ **c** Mg_3N_2.

2 Some compounds contain molecules. Suggest and draw molecules for the following compounds.

a SO_2 **b** CH_4 **c** C_2H_6 **d** N_2O.

3 Look back at the experiment carried out by Craig's class.

a What was their input variable?
b What was their outcome variable?
c How did they work out the mass of oxygen that had reacted?
d Suggest two reasons why the mass of oxygen reacting with 3 g of magnesium was low.

G1 Rock breaking

What are rocks?

Rocks are everywhere. We walk on them and build with them. There are many types of rock. Some are made of **grains**. Some are made of **crystals**. The grains and crystals are made of compounds called **minerals**. Different rocks are made up of different minerals or different mixtures of minerals.

A close fit?

The way the grains or crystals fit together gives a rock its **texture**.

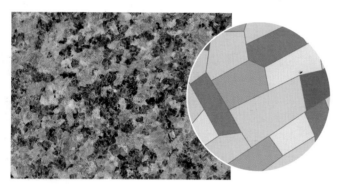

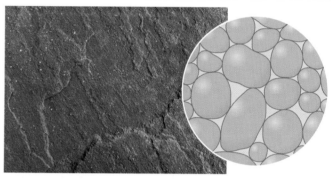

Look at the photo above of **granite**. It is made of crystals which fit together with no gaps between them. We say they are **interlocking**.

Look at the photo above of **sandstone**. It is made of grains which do not fit together. There are gaps between them because they have a round shape. They are **non-interlocking**.

What happens if you drop water onto the surface of a rock? It stays on the surface of some rocks, but with others it soaks into them. Rocks with non-interlocking grains are often **porous**. Water can get into the gaps between the grains and soak in.

a Granite has interlocking crystals. What will happen if you drop water onto granite?

Weathering

All rocks gradually get broken down into smaller pieces where they are. We call this **weathering**.

Look at the photo on the right. The small rocks at the bottom of the slope used to be part of the mountain. They broke off the mountain and fell to the ground. Rocks can fall at any time without warning, making it dangerous for mountaineers in the area.

When rock is broken into smaller pieces, but not changed into different substances, we call it **physical weathering**. This is a physical change – no new substances are formed. It can be caused by water and changes in temperature.

Water and ice

In the mountains, there is often a lot of water which freezes. The diagrams on the right show what happens when water gets into small cracks or holes in a rock. This is called **freeze–thaw weathering**.

Some rocks are more resistant to weathering than others. Porous rocks with gaps between the grains and cracks in them are more easily freeze–thaw weathered than non-porous rocks.

ⓑ Explain how water can break a big rock into small chunks. Use the word 'ice' in your explanation.

ⓒ Why is granite less likely to be freeze–thaw weathered than sandstone?

1 Water gets into cracks in the rock.

2 The water freezes and it expands.

3 The force of the ice makes the crack bigger.

4 The crack gets so big that part of the rock breaks off.

Hot and cold

In a desert, it is very hot during the day, but it can get very cold at night. During the day, the hot rocks expand. During the cold nights, they contract.

Rocks are a mixture of different minerals, and some of them expand and contract more than others. This causes huge forces of strain in the rock.

This expansion and contraction happens every day and night, causing cracks to appear in the rock. Eventually, the rock breaks apart. The same thing happens to the smaller pieces. This happens until all the rocks are broken down to small grains of sand.

ⓓ Special bricks are needed to line an oven used for firing pots. The oven is heated to high temperatures and cooled down again. What special properties do you think the bricks need to have?

expansion

contraction

stresses in the rock cause it to crack

Plants

Plant roots can force their way through cracks in rocks. When they grow they make the cracks bigger. This is called **biological weathering**.

Questions

1 Give three ways that rocks can be physically weathered.

2 Explain, in detail, why deserts are full of sand.

3 Explain how you think tree roots could break down rocks.

4 Some physical weathering needs water or extreme temperature changes. Suggest what types of climate would produce conditions for:

   a fast physical weathering   **b** slow physical weathering.

For your notes:

● Rocks are made up of **grains** or **crystals**. These are made of different compounds called **minerals**.

● **Physical weathering** happens when rocks are broken down into smaller pieces but not changed into different substances.

● Physical weathering is caused by frost and changes in temperature.

G2 Disappearing rocks

Acid on the rocks

Many statues and buildings are made of a rock called **limestone**. Limestone is weathered easily by rainwater. Rainwater is a slightly acidic solution. This is because it contains carbon dioxide from the air dissolved in it.

The photo on the right shows some rainwater being tested using a pH probe. The probe gives a reading of pH 5.5.

a If the rainwater was more acidic, suggest what its pH might be.

b Give two reasons why the pH probe is more precise than using universal indicator paper.

A limestone statue.

When rainwater falls on rocks, it can get into any cracks or into the gaps between grains.

Limestone is made of calcium carbonate. When acid reacts with calcium carbonate, a new substance is made which is soluble in water, so it dissolves. Carbon dioxide is also produced. Over thousands of years, holes and cracks appear, as shown on the left in the photo of limestone pavement at Malham Cove in Yorkshire.

The limestone is completely changed by the acid and new substances are produced. Many cliffs, such as those at Lyme Regis, are made from limestone and are gradually disappearing.

c Write a general word equation for the reaction between an acid and a carbonate.

Weathering, when chemical changes take place in the rock, is called **chemical weathering**. Chemicals in rainwater react with minerals in the rock and produce new substances.

In the photo above, chemical weathering has changed the colour of the stone.

Over millions of years, rainwater can dissolve limestone underground to make caves. Grotte Bournillon in France, shown in the photo on the right, has a cave mouth 80 m high, the largest cave entrance in Europe.

Making comparisons

Granite is weathered more slowly than limestone because it contains different minerals. Some of the minerals in granite react slowly with acid to make new substances. Some of the substances formed are soluble in water, so they dissolve and are carried away by the water. This gradually weakens the rock, causing it to crumble and change. Some minerals in granite are resistant to chemical weathering.

Time to experiment

Tia and Danielle added acid to samples of limestone and granite. The limestone fizzed, giving off a gas. There was no reaction with the granite.

granite

limestone

d (i) **Explain their results.**
(ii) **Which rock would be affected most by chemical weathering?**
(iii) **Compare limestone and granite. What are the advantages and disadvantages of using these two materials for building a new town hall?**
(iv) **Suggest how Tia and Danielle could modify their experiment to collect the gas given off by the limestone and acid. Draw a diagram of the apparatus they might use.**
(v) **How could they test the gas to prove that it was carbon dioxide?**

Tia and Danielle's experiment.

Soil

When rock is chemically and physically weathered, it turns into soil. The top layer is called **topsoil**. This contains tiny grains of sand and clay. Larger pieces of rock in the soil are stones. Topsoil also contains decayed plant and animal matter called **humus**. Humus is rich in minerals that plants need for growth.

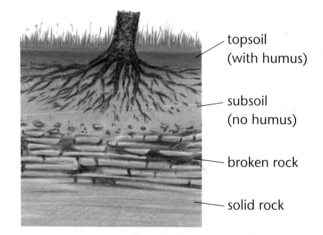

topsoil (with humus)

subsoil (no humus)

broken rock

solid rock

Questions

1 Explain the difference between chemical weathering and physical weathering.

2 A gravestone from 1780 is very difficult to read. The letters on a gravestone from 1945 are much clearer. Explain why you think this might be.

3 Chemical weathering needs hot, damp conditions. Suggest two types of climate where chemical weathering would not happen very much.

4 What is soil made up of? How is it formed?

For your notes:

- Rainwater becomes acidic when carbon dioxide dissolved in it.

- **Chemical weathering** is caused by reactions between acidic rainwater and minerals in the rock.

- Soil forms when rocks are weathered.

G3 Transporting rock

Where did it go?

When this house in the photo was built, it was a long way from the edge of the cliff. Over the years the rock has collapsed into the sea. Finally, the edge of the cliff is so close that the house is in danger of collapsing too.

The weathered fragments of rock from the cliff are carried away by the movement of the waves and currents in the sea. As they are carried, the rocks knock against each other and bits break off. The fragments become more rounded and smaller. The broken pieces of rock wear away the solid rocks of the cliff when waves crash against it. These processes are called **erosion**. Erosion can be caused by moving water (seas or rivers), the wind or glaciers.

Water

Rivers can carry away large amounts of weathered rock from mountains and hills. The size of rock fragments that the river can carry depends on the speed of the water. Rivers move very fast near to their source in the mountain. Here the water can carry quite large fragments of rock. Rivers move more slowly when they get close to the sea and so can only carry small grains of rock. The sea can also carry weathered material away.

sand carried by the flow

fast

stones roll and bounce

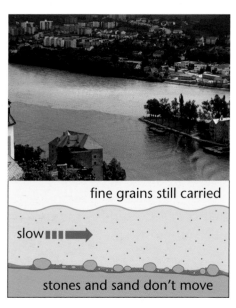

fine grains still carried

slow

stones and sand don't move

a Would erosion be quicker near the source of a river or near where it reaches the sea? Explain your answer.

Wind

The wind can only carry small, light grains of sand. It cannot carry large fragments of rock because they are too heavy. If the wind is strong and the air is moving very quickly, it may be able to carry slightly larger grains of sand.

b Explain the relationship between the speed of air movement and the size of grains the wind can carry.

Something tells me that fossil must have got there because the Earth has changed …

Mike had a gut reaction. He was thinking **intuitively**.

I must read about fossils in my Beginner's Guide to Geology.

Mike was checking the facts. He was thinking neutrally.

We should find out if there are any more of these fossils and make a note of where they are found.

Mike was making notes to help others. He was thinking **constructively**.

I mustn't rush into saying I have found something new – I might be mistaken.

Perhaps I should decide which of my ideas to follow.

Mike was thinking about his thinking. He was **reflecting**.

Mike was thinking about the risks. He was thinking cautiously.

Lateral thinking

Sometimes things we see do not fit in with what we already know. We need to think of new ideas and look at other scientific facts. This is called **lateral thinking**. Lateral means sideways, so thinking laterally means thinking in a different direction.

Michelle suddenly had a flash of inspiration.

You are right, the oldest fossils are found in the deepest rocks, but Earth movements or earthquakes can push rocks up. Maybe Earth movements raised up the rocks, together with any fossils in them.

c **What new scientific fact did Michelle take into account?**

Famous scientists

Famous scientists think about problems in different ways. They often need to think laterally to explain things.

On the right are some quotations about Mary Anning.

d **Which of these quotations do you think suggests that Mary Anning could think laterally? Explain your answer.**

'All the professors and clever geologists acknowledge that she understands more of the science than anyone else in the kingdom.'

'A being of imagination – she has so many ideas and such power of communicating them.'

'She is a history and a mystery.'

'A very clever, funny creature.'

Questions

1 How many different types of thinking did Mike and Michelle use to solve the problem?

2 Which type of thinking explained how the fossil of the fish that lived 190 million years ago got up into the cliff face?

3 Imagine you found a fossil of a dinosaur bone on the beach. Explain how you think it could have got there.

H1 Hard rocks

Rock solid

Grains from weathered and eroded rocks build up in sedimentary layers. Over millions of years these sedimentary layers gradually turn to **sedimentary rocks**. The rocks sometimes contain fossils from dead remains of creatures in the sedimentary layers.

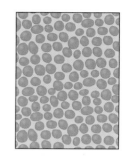
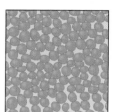

How do the grains turn to solid rock? As the sediment builds up, the weight of the layers above causes enormous pressure on the lower layers. The grains become pressed tightly together. This is called **compaction**. Between the grains are tiny gaps that contain water.

In the lake or sea where sediments are deposited, there are chemicals dissolved in the water. The compaction squeezes out most of the water from the tiny gaps between the grains of sediment. The chemicals in the water crystallise in the gaps between the grains. They 'glue' the grains together. This is called **cementation**.

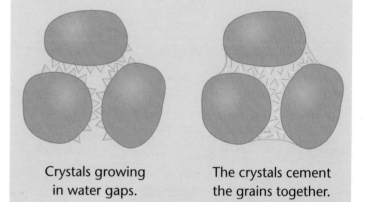

Crystals growing in water gaps.

The crystals cement the grains together.

a How are the grains in sedimentary rock stuck together?

Types of sedimentary rock

There are many types of sedimentary rock, made from different minerals. Examples of sedimentary rocks include limestone, sandstone and shale.

Limestone is mostly made from the shells and bones of small sea animals. Millions of years ago there were warm shallow seas full of sea animals. Their remains were deposited on the sea bed. The calcium carbonate in the rock comes from their shells and bones.

Limestone occurs in different colours. This is because different compounds mix with the calcium carbonate. Yellow or brown limestones get their colour from iron compounds.

Do you remember?

Limestone is made of calcium carbonate. When you add acid to calcium carbonate, it fizzes and gives off carbon dioxide.

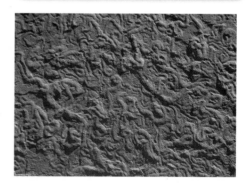

These three rocks are all different limestones.

Limestone is a very useful rock. We use it for building roads and buildings, and it is also found in toothpaste and make-up. We even add it to bread to give us calcium!

Sandstone is made from grains of sand. A similar rock called gritstone is made from angular grains of grit. Sandstone and gritstone are also used as building materials. Gritstone is hard and was used to make millstones for grinding corn into flour.

Shale and clay are made from very fine grains of rock. There is so little space between the grains that these rocks will not let water pass through.

Sandstone.   *Shale.*

Changing rocks

When sedimentary rocks are buried deeper in the Earth, they are squeezed more and more. They become heated or put under high pressure, or both. The minerals in the rock change chemically without the rocks actually melting. New minerals form as crystals. The new rock is called **metamorphic rock**. The word *metamorphic* means 'changed form' in Greek.

In metamorphic rocks, the crystals are interlocking so the rock is not porous. Metamorphic rocks do not usually have fossils because the fossils are destroyed when the rock changes.

Marble is a metamorphic rock with an attractive sugary texture. Marble usually looks white or grey but sometimes it has coloured patches. It is often used as a building material or for sculptures. Marble reacts with acid.

Shale contains minerals in layers like plates. Under pressure newly made crystals line up in bands to make **slate**. Slate splits easily between the layers and is used to make roof slates. Slate does not react with acid.

Limestone (left) is changed into marble (right) by heat and increased pressure.

b Look at the photos above. Describe the differences between limestone and marble.

c A metamorphic rock made from a sedimentary rock with a lot of heat and pressure will not have any fossils. Why do you think this is?

Shale (left) is a sedimentary rock. When compressed and heated it changes to slate.

Questions

1 Explain how sedimentary rocks are formed, using the words compaction and cementation.

2 Limestone is mainly calcium carbonate, which is white. Why do we find different coloured limestones?

3 Why do sedimentary rocks often contain fossils?

4 Explain the difference between a sedimentary rock and a metamorphic rock.

5 What chemical test would you use to show that the composition of marble is different from that of slate?

6 Describe the uses of marble and slate. Why do you think these rocks are chosen for these uses?

For your notes:

- **Sedimentary rocks** are made from layers of sediment. The pressure of the layers causes **compaction** and **cementation** of the grains.

- **Limestone**, **sandstone** and **shale** are all sedimentary rocks.

- **Metamorphic rocks** are made when heat or high pressure or both changes existing rocks.

- **Marble** and **slate** are metamorphic rocks.

Igneous rocks

As well as sedimentary and metamorphic rocks, there is a third type of rock called **igneous rock**.

Igneous rocks are made from liquid or molten rock called **magma**. Magma comes from inside the Earth where it is very hot. The magma rises up to the crust of the Earth. Igneous comes from the Greek word for 'fire'. Igneous rocks are hard rocks made of crystals. We say they are **crystalline**.

When the magma reaches the Earth's surface it is called **lava**. Sometimes the magma is forced out and erupts from a **volcano**, as shown in the photo below right. Magma might also be blown out as volcanic ash.

Types of igneous rock

At the Earth's surface, lava cools quickly and forms **extrusive** igneous rock with small crystals. **Basalt** is an example of this sort of rock. Sometimes the magma never reaches the surface but cools slowly underground. This forms **intrusive** igneous rock with large crystals. **Granite** is an example of this sort of rock.

solid crust

hot rock   molten core

A slice through the Earth.

Molten rock cooling quickly forms basalt.

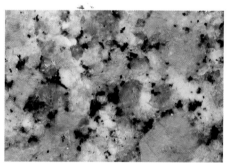

Molten rock cooling more slowly forms granite.

Fast and slow cooling

Class 8T used a substance called salol to model the effect that cooling quickly or cooling slowly has on the size of crystals that form. They melted the salol in a test tube and then poured a few drops onto a cold microscope slide. They repeated this using a warm microscope slide.

The photos show that on the cold slide (left), the salol cooled quickly and the crystals were small. On the warm slide (right), the salol cooled slowly and the crystals were bigger.

Later, the class imagined that they were particles in molten rock. They moved around the classroom shaking hands as often as possible with each other. When the teacher said 'Start cooling', they stuck together in groups.

ⓐ How could this activity show that slow cooling gives larger crystals? Explain your answer using the idea of particles.

Differences between igneous rocks

Different types of igneous rock have different colours because they contain different minerals. We can classify igneous rocks into two groups. The lighter coloured rocks contain minerals which are rich in silica but not much iron. The darker coloured rocks contain less silica and a lot of iron-rich minerals.

Go with the flow

Volcanoes with silica-rich rocks produce thick lava that flows slowly.

Volcanoes with iron-rich rocks and less silica erupt frequently. They produce thinner basalt lava that flows quickly.

ⓑ Look at the photos on page 78 of granite and basalt. Which rock contains more iron-rich minerals?

Different densities

To find the **density** of a rock, you need to know the mass of the rock sample and its volume.

- Find the mass of the rock in grams using a balance.

- Attach a piece of string to the rock and lower it into a Eureka can full of water. The rock will push out its own volume of water through the spout.

- Measure the volume of the water.

Look at the diagram on the right. This rock has a mass of 30 g and a volume of $10 \, cm^3$. So a volume of just $1 \, cm^3$ of the rock would have a mass of 3 g. The density of the rock is $3 \, g/cm^3$.

Some igneous rocks have a higher density than others. Rocks containing more iron have higher densities than rocks containing more silica.

ⓓ Which rock will have the higher density, granite or basalt?

ⓒ How does the chemical composition of the magma lead to different types of volcanic activity?

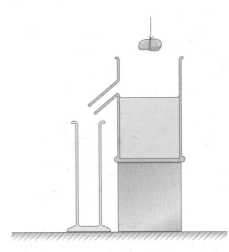

For your notes:

- **Igneous rocks** crystallise from molten rock called **magma**.

- Igneous rocks with small crystals cooled quickly at the Earth's surface. Igneous rocks with large crystals cooled slowly underground.

- Igneous rocks contain different minerals. This gives them different colours. They also have a different **density**.

Questions

1 Explain how igneous rocks are formed.

2 A geologist found an igneous rock that has small crystals. Did it cool close to the surface or several kilometres deep? Explain your answer.

3 a Give two properties that are similar for granite and basalt.
  b Give two properties that are different for granite and basalt.
  c Explain how these different properties came about.

H3 Rock on

The rock cycle

Rocks in one of the three main groups may change into another group over millions of years. This is called the **rock cycle** and it is shown in the diagram below.

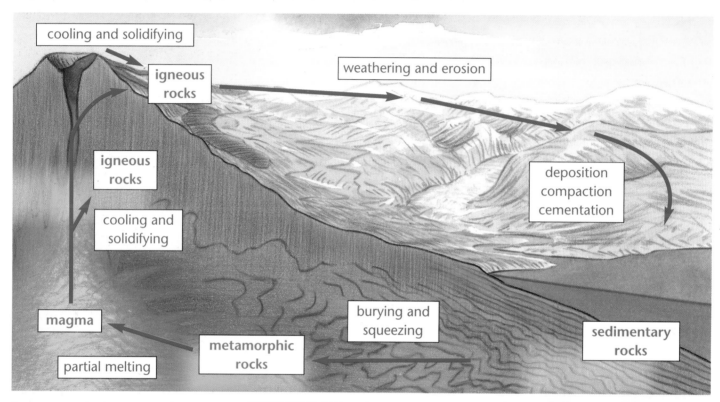

cooling and solidifying

igneous rocks

weathering and erosion

igneous rocks

deposition compaction cementation

cooling and solidifying

magma

burying and squeezing

sedimentary rocks

partial melting

metamorphic rocks

Igneous rocks are formed when magma cools and solidifies. These rocks are worn away by weathering on the Earth's surface. The weathered rocks are carried away by rivers and eroded. Rock fragments are deposited as sediment on the sea bed to make sedimentary rocks.

Some sedimentary rocks are changed by heat and pressure into metamorphic rocks. Some metamorphic rocks partially melt and produce magma.

Which type?

The properties of the rocks you have met in this unit are summarised in these photos. These include grain or crystal size, hardness, texture, colour and chemical reactions.

Limestone: a pale, hard, often porous rock. It often contains fossils. Limestone reacts with acid.

Sandstone: a yellow, often porous rock made of grains of sand. Most sandstones do not react with acid.

Shale: a soft, grey, porous rock. It has very clear layers and often has fossils in it. Most shales do not react with acid.

Slate: a hard, grey, green or purple non-porous rock. It is easily split into layers. Fossils are not found in slate and it does not react with acid.

Marble: a hard, non-porous rock that is normally white. It is made of crystals and usually has no fossils or layers. It reacts with acid.

Basalt: a hard, dark, non-porous rock with very small crystals. There are no layers and it does not react with acid.

Granite: a hard, non-porous rock with large crystals. It looks mottled as it is made of three coloured minerals. There are no layers and it does not react with acid.

a List the seven rocks shown, and say whether each is igneous, sedimentary or metamorphic.

b How can you tell the difference between shale and slate?

Changing Earth

The Earth's surface is continually changing. The rock cycle makes and breaks rocks. Many of the rocks near the Earth's surface are useful as building materials or contain useful minerals such as copper. We dig a large hole or mine to get out the rocks. Mines can look ugly.

c Mining provides us with useful rocks and minerals, but what disadvantage does it have?

Questions

1 Write the following parts of the rock cycle in the correct order, starting with weathering and erosion.

    weathering and erosion   **deposition**   **rocks melt**
    sedimentary rocks form   **burying and squeezing**
volcanic eruption   **metamorphic rocks form**   **igneous rocks form**

2 You could use a sandwich as a model for the formation of sedimentary rocks.

  a What could you do to the sandwich to model the metamorphic process?

  b Can you think of something to model igneous rock formation?

3 As well as being buried in the Earth, sedimentary rocks can also be uplifted by earth movements such as earthquakes. Slate or sandstone containing fossil sea shells is sometimes found on the tops of mountains. Explain in terms of the rock cycle how this is possible.

4 Imagine you are a pebble on a beach. Prepare a poster to describe the story of your life.

For your notes:

- The three main groups of rock, igneous, sedimentary and metamorphic, are made in different ways and have different properties.

- They may change from one group to another over millions of years.

- These changes are summarised in the **rock cycle**.

H4 Name that rock

Why and what?

These words are all question words.
Questions are very important in science:

Why? *What?* *How?* *Which?* *Where?*

● Scientists carry out investigations to find out the answers to questions.

● It is important to ask the right kinds of questions if you want to find out anything useful.

● Scientists use questions when they are classifying things and putting them into groups. A **key** is a set of questions that helps us to classify things.

Out in the field

Class 8T are on a geology field trip. They are examining the rocks around them and testing them with acid. They are collecting small samples of different rocks to take back to the classroom. There they will find out about the properties of each rock, so they can look up in a reference book what sort of rock it is. This will also tell them how it was formed and help them discover how the landscape around them was made.

a To help the class identify the rocks back at the classroom, write a list of the questions they will need to ask and answer about each sample.

b In your group, look at all the questions you have written down. Decide which questions will be the most useful when identifying rocks.

c Is there something that all the useful questions have in common?

Identifying the rocks

Look at these photos of the rocks.

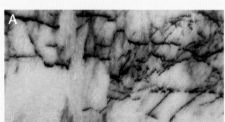
A

B

C

d Use your questions to help you to identify each of the three rocks.

Back at the classroom, Lisa's group found they hadn't kept careful enough notes about their samples. To make matters worse, Lisa had felt ill by the end of the field trip and had gone straight home and taken their rock samples with her. The rest of the group put together the notes they had. These are shown on the next page.

e Look at the information Lisa's group brought back. Think about what you know about the different properties of rocks. In your group, write down the questions you would need to answer in order to identify their five rocks.

f Look again at the type of questions you need to ask. Are they similar to the useful ones you identified earlier?

Types of question

You may have found that the most useful questions are the ones that you answer with either 'yes' or 'no'. Scientists use these kinds of questions to make keys. We use keys to identify things and classify them, such as rocks, elements, plants and animals. These questions are called **closed questions** because the type of answer you can give is closed down to 'yes' or 'no'.

In contrast, an **open question** can have several possible answers. For example, 'What is the appearance of the rock?' is an open question.

g Look at these questions. In your group, decide if each one is an open or a closed question.

1. The rock has large and small pebbles in it. The pebbles are smooth and rounded. There are bits of sandy grains in between.

2. Pale orange rock. Quite large grains. May be small bits of shells.

3. Several different coloured bits in the rock. Doesn't react with acid.

4. Contains large black flat crystals which are lined up. Some pale crystals. Very hard.

5. Dark grey/black rock all over. Doesn't react with acid.

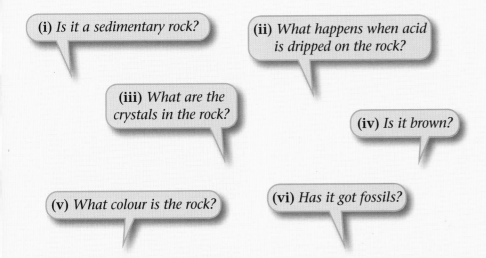

(i) Is it a sedimentary rock?

(ii) What happens when acid is dripped on the rock?

(iii) What are the crystals in the rock?

(iv) Is it brown?

(v) What colour is the rock?

(vi) Has it got fossils?

Questions

1 Write a key to classify the three main groups of rock. Use some of the questions you have written and what you know about the properties of the three main groups of rock.

2 Discuss what kind of question is best for classifying things. Explain your answer.

3 Imagine you are going into the field to make observations and collect information about rocks. Describe how you would organise your records to make sure you come back with all the information you need to identify the rocks. Explain why you would organise your records in the way you suggest. (*Hint:* think about using ICT.)

I1 What temperature?

Temperature

It's hot in here!

It's cold in here!

The hall feels hot to Stephen and cold to Miriam. What is hot to one person may be cold to another. We can't depend on our bodies to tell us how hot things are.

We decide how hot something is by measuring its **temperature** in **degrees Celsius**. On the Celsius scale, freezing water is at 0°C and boiling water is at 100°C. We show that a temperature is being measured in degrees Celsius by putting °C after it. Anders Celsius came up with the idea of having 100 degrees between freezing water and boiling water in 1742.

Did you know?

There are many other temperature scales. On the kelvin scale freezing water is 273 K and boiling water is 373 K. The kelvin scale is never negative, however cold it is. In the USA they use the Fahrenheit scale. 0°C is the same as 32°F.

Measuring temperature

We use a variety of different thermometers to measure temperature. The photos here show people using glass thermometers, temperature probes, digital thermometers and temperature-sensitive plastic strips. Other temperatures, like the temperature of the core of the Sun, have to be estimated from other observations. The table on page 85 and the photos show the temperature of many different objects and places.

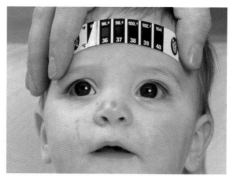

The temperature of the human body is 37°C.

Water freezes and melts at 0°C.

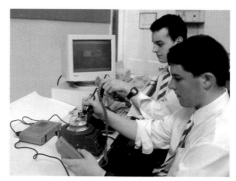

Water boils at 100°C, ethanol boils at 79°C.

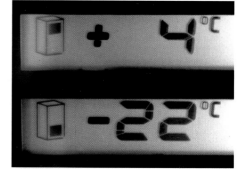

We keep fridges at 4°C and freezers at −22°C.

Environment, object or event	Temperature in °C
'Room' temperature (comfortable temperature for humans)	20
Coldest place in Solar System (Triton, moon of Neptune)	−235
Sun's core	15 400 000 000
Lowest possible temperature	−273
Temperature at which glass softens	820
Lightning	30 000
Bunsen burner flame	1500

a Use the information in this table and the photos on page 84 to answer these questions.

(i) Why couldn't you use a glass thermometer in a Bunsen burner flame?

(ii) Work out the temperature change if you heat ethanol from room temperature to boiling.

(iii) Work out the temperature change if you make ice in a freezer.

Hot, not cold

Stephen is right. There is no such thing as 'cold'. To understand about 'hot' and 'cold' we need to talk about **energy**. Hot things contain a lot of **thermal** (heat) **energy**. Cold things contain less thermal energy.

Imagine Stephen touching something hot, like a radiator. The thermal energy moves from the radiator, which is hotter, to his hand, which is colder. Stephen decides the radiator is hot because the thermal energy flows into his hand.

Shut the door, don't let the cold in.

There's no such thing as cold.

Shut the door. Don't let the heat out!

Imagine Stephen touching something cold, like an ice cube. The thermal energy moves from his hand, which is hotter, to the ice cube, which is colder. Stephen decides that the ice cube is cold because thermal energy leaves his hand.

b Why did Miriam feel cold when she walked into the hall from the living room?

If you study the table you will see that there is a lowest possible temperature, −273 °C, but no highest possible temperature.

Cold is the absence of energy. As you take energy away from an object, it gets colder. Once you have taken all the energy away, the object is as cold as it can be. Scientists call this temperature **absolute zero**.

You can always add more energy. There is no upper limit on temperature.

Questions

1 Describe the Celsius scale for measuring temperature.

2 Why do we need thermometers?

3 Explain why 'letting the cold in' is an unscientific thing to say.

4 Design a diagram to show all the temperatures mentioned in the table above. Your diagram should contain one or more scales to show how the temperatures relate to each other.

For your notes:

- **Temperature** is measured in **degrees Celsius, °C**.

- Freezing water is 0 °C. Boiling water is 100 °C.

- Cold is the absence of **thermal energy**.

I2 Temperature and energy

I2 Temperature and energy

Learn about:
- Temperature and thermal energy

How long?

Stephen makes a cup of coffee. He only fills the kettle to the 'two cup' mark. The kettle takes just under a minute to boil.

Later, everyone wants a hot drink. Stephen fills the kettle to the 'six cups' mark. The kettle takes just over three minutes to boil.

The temperature increase is the same, but the kettle took longer to boil. There was more water, so more thermal energy was needed to get the same temperature.

a The tap water was at 20 °C. What was the temperature rise both times?

b How much more energy was needed to heat the water the second time?

Do you remember?

Water, like all materials, is made up of particles. Read about particles on page ix.

Explaining heating

We can use a 'money and children' model to discuss what happens when different amounts of water are heated.

Look at the cartoons and read the captions. Both kettles are heated to 80 °C. In the model, each child has to be given 6p.

c How much money is given to:
(i) the part-full kettle?
(ii) the full kettle?

These 20 children represent the particles of water in the part-filled kettle. Each child has 2p to represent the starting temperature, which is 20 °C.

These 100 children represent the particles of water in the full kettle. Each child has 2p to represent the starting temperature, which is 20 °C.

Temperature and energy

You can think of the temperature of a material as being the 'energy per particle'. If there is more energy per particle, the temperature is higher. If there is less energy per particle, the temperature is lower. Look at the two cartoons on the left. In the first there is 5p per child, so the temperature is higher than in the second, where there is 2p per child, so the 'temperature' is lower.

86

Look at the first cartoon again. There is a total of 10p. In the second cartoon there is a total of 12p. The temperature is higher, but the energy is lower.

When we think about energy we have to think about how many particles there are, as well as the temperature. Hotter does not always mean more energy.

Strange thoughts

There is more thermal energy in a tub of bath water …

There is more thermal energy in a wrought-iron gate …

There is more thermal energy in an iceberg …

… than in a kettle of boiling water.

… than in a red-hot poker.

… than in a vat of molten iron

d Think again about the part-full and full kettles.
 (i) Which would be hotter after one minute of heating?
 (ii) Which would contain more energy when it was boiling?

e Which would have more energy:
 (i) 1 kg of lead at 400 °C or 1 kg of lead at 40 °C?
 (ii) 1 g of ice at −20 °C or 1 g ice at −10 °C?
 (iii) 1 g of ice at −10 °C or 100 g ice at −10 °C?

Questions

1 Look at the 'Strange thoughts' section on this page. Explain in your own words how a red-hot poker made of iron can contain less energy than a wrought-iron gate.

2 In two experiments, scientists mixed hot iron with cold water and measured the final temperature.

Hot iron	Cold water	Final temperature of mixture
1 cm³ iron at 1000 °C	1000 cm³ water at 20 °C	28 °C
8 cm³ iron at 1000 °C	1000 cm³ water at 20 °C	84 °C

(i) Which contained more energy, 1 cm³ iron at 1000 °C or 8 cm³ iron at 1000 °C? Explain your answer.

(ii) Explain, using the 'children and money' model, why the mixture's final temperature was hotter in the second experiment.

For your notes:

● Temperature is the energy per particle.

● Smaller, hotter things can have less energy than colder, larger things.

13 Bigger and smaller

Allow for it

Railway tracks are made with gaps. The gaps allow the tracks to get bigger on a hot day without bending out of shape.

We say that materials **expand** when they are heated.

The cables in the photo on the right are made of aluminium. The cables are not tight between the pylons. This allows them to get shorter in the winter.

We say that materials **contract** when they are cooled.

a Explain why there are gaps between the concrete sections of a motorway.

Explaining expansion

Why do solids expand when they are heated? To understand this, we have to think about the particles.

lowest possible temperature

warmer

still warmer

As the particles vibrate more, they take up more space. Each particle stays the same size but it vibrates more, so it pushes against the particles next to it. The solid expands. This is shown in the diagram on the left, the solid on the right is larger. When the solid cools and energy is removed, the particles vibrate less. The solid contracts.

b Compare the amount the particles vibrate in a hotter solid and in a colder solid.

Liquids

Liquids also expand when they are heated and contract when they are cooled. This is because the particles vibrate more at higher temperatures, and slide over each other more quickly. The diagram on the left shows the same number of particles in a cooler and a hotter liquid. The particles in the hotter liquid take up more space.

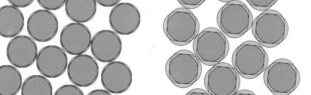
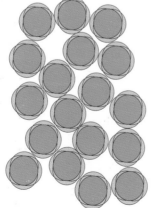

cooler liquid   hotter liquid

c Explain why liquids contract when cooled.

Gases

When gases are heated, the particles move faster. When the gas is trapped in a container that can change shape, like a balloon, the particles hit the sides of the balloon more often and harder. The balloon is stretched further, increasing its volume. The gas then fills the extra space. This means that the particles are further apart, with more space between them.

d Explain why an inflated balloon gets smaller when it cools.

Density changes

Some materials, like lead, are **dense**. Other materials, like cork, are less dense. Density is the amount of mass in a certain amount of volume. You work out the density using this formula:

$$\text{density} = \frac{\text{mass}}{\text{volume}}$$

Density is measured in 'grams per centimetre cubed', g/cm^3, or 'kilograms per metre cubed', kg/m^3. Water has a density of $1\,g/cm^3$.

When materials expand they keep the same mass – the number of particles has not changed so the mass stays the same. However, the volume has increased. What happens to the density?

Imagine $4\,g$ of a material. It has a volume of $4\,cm^3$. Its density is $4 \div 4 = 1\,g/cm^3$. If you heat the solid, it expands. Its mass is still $4\,g$. Its volume is now $5\,cm^3$. Its density is now $4 \div 5 = 0.8\,g/cm^3$. The material is less dense.

When materials expand, their density decreases. When materials contract, their density increases.

Do you remember?

The particles in a gas are far apart and are moving quickly.

For your notes:

- When you heat a solid or a liquid it **expands** because the particles vibrate more and take up more space.

- When you cool a solid or liquid it **contracts** because the particles vibrate less and take up less space.

- Gases can also expand when heated, increasing the spaces between the particles. Gases contract when cooled, decreasing the spaces between the particles.

- When a material expands, it becomes less **dense** because there are the same number of particles in a larger volume.

Questions

1 Rivets are used to hold pieces of metal together. The rivets are heated up, then put through holes in the metal. The end is then flattened off.

   a What will happen to the rivet as it cools down?
   b What happens to the metal sheets as the rivet cools down?

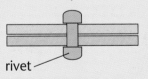

rivet

2 Use your understanding of the particle model to explain why:

   a gases are much less dense than solids or liquids.
   b the density of solids, liquids and gases usually decrease when the temperature increases?

3 Some solids expand more than others for the same rise in temperature. For example, iron expands more than glass. Explain why running hot water over a jam-jar lid that's stuck tight will help to release it.

I4 All change

Watching ice melt

Darren and Jackie are investigating changes in state by heating ice from the freezer. The apparatus they use is shown in the diagram below. The graph of their results is also shown.

> **Do you remember?**
>
> The three **states of matter** are solid, liquid and gas. **Changes in state** are melting, freezing, boiling and condensing.

That's weird. It's got flat bits. Perhaps the equipment stopped working.

Look, the flat bits are at 0 °C and 100 °C. It goes flat during melting and boiling.

Why does the temperature stay steady during melting and boiling? You need to think very carefully about the particles.

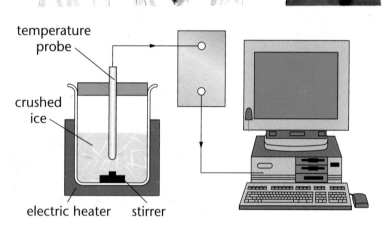

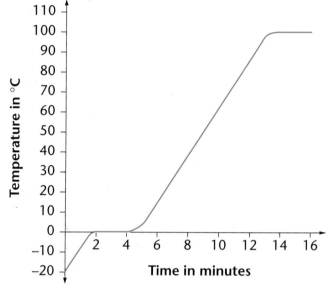

Melting

Look at the diagram on the right showing the particles in a solid. The yellow particle is touching six other particles. There are six **forces of attraction**, shown by the black stars, between the yellow particle and other particles. These forces of attraction add up to hold the particles together strongly in a solid. Look at the yellow particle in the liquid. It is only touching three other particles. The forces of attraction hold the particles together less strongly in a liquid.

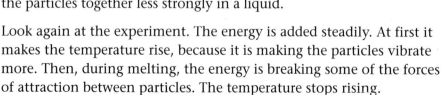

solid   liquid

Look again at the experiment. The energy is added steadily. At first it makes the temperature rise, because it is making the particles vibrate more. Then, during melting, the energy is breaking some of the forces of attraction between particles. The temperature stops rising.

When enough of the forces of attraction are broken, the particles can slide over each other, and the solid has become a liquid. Then the energy goes into making the particles vibrate more, the temperature increases and the graph begins to rise.

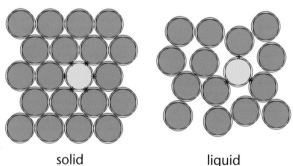

a Explain why:
  (i) the temperature rises when you put more energy into the ice
  (ii) the temperature does not rise while the ice is melting.

Boiling

Look at the graph again on page 90. At 100 °C, while the liquid is boiling, there is no change in temperature. During boiling, the energy is breaking the forces of attraction between the particles in the liquid, rather than increasing the temperature.

When all the forces of attraction are broken, the particles are free to fly around the room as part of a gas.

b Explain why there is no temperature rise while the liquid boils using the words below.

energy   forces of attraction   particles

Different directions

Both melting and boiling are reversible changes. If you remove energy from steam it cools down, then condenses, then cools down again and freezes. During condensation and freezing the temperature does not fall, even though energy is being removed.

Energy was put in to break the forces of attraction between particles during melting and boiling, so energy must be given out when new forces of attraction are made during condensing and freezing.

c Sketch a graph showing the temperature change as you remove energy from steam.

d Think about removing energy from steam.
 (i) Suggest a temperature at which the energy removed is making the particles move less and so the temperature is dropping.
 (ii) Suggest a temperature at which the energy is given out as new forces of attraction are made and the temperature stays constant.

Questions

1 Where does the energy go when you:

 a heat water from 20 °C to 40 °C?   **b** boil water at 100 °C?

2 Sketch a graph showing Darren and Jackie's results.

 a (i) Label the melting point.   (ii) Label the boiling point.
 b Label where there is:
  (i) only solid   (ii) only liquid
  (iii) liquid and gas together   (iv) solid and liquid together.
 c Put a star where energy is being transferred into the substance to break forces of attraction.

3 Explain carefully why the temperature of water in a domestic freezer stays at 0 °C until all the liquid has frozen, and then gradually drops to −20 °C using the words below.

particles   energy   vibration   temperature   forces of attraction

Did you know?

Steam at 100 °C causes worse burns than water at 100 °C. As the steam condenses it gives out energy, making the burn worse.

For your notes:

- The **changes of state** are melting, boiling, condensing and freezing.

- Melting and boiling happen when energy is put in. Condensation and freezing happen when energy is taken out.

- During changes in state the temperature of a substance stays the same.

- During melting and boiling, energy is needed to break the **forces of attraction** between particles.

- During condensation and freezing, energy is given out when new forces of attraction are made between particles.

15 Conduction

Learn about:
- Thermal energy transfer in solids
- Conductors and insulators

Heat it up

Stephen's dad forgot to take the chicken out the freezer. Dinner is going to be very late unless they can quickly defrost the chicken. It has to be totally defrosted before they can cook it. They need to transfer thermal energy from the surroundings to the chicken.

At first, the surroundings are much hotter than the chicken. Thermal energy is transferred from the surroundings to the chicken. The surroundings cool a little and the chicken heats up. Finally, everything ends up at the same temperature.

a Explain why the surroundings only cool by a few degrees Celsius, while the chicken warms up by about 40°C.

Conduction in solids

In solids, like the chicken, thermal energy is transferred from the hotter part to the cooler part by **conduction**. The thermal energy is passed from particle to particle without the particles moving from their places in the material.

Some of the particles in the chicken are shown in the diagram. The surface of the chicken is hotter than the rest of it. Where the temperature is higher, the particles have more energy and vibrate more. These particles hit against the neighbouring particles, making them vibrate more and thus increasing the temperature of that part of the solid.

The energy is passed from particle to particle through the chicken below, until all the particles are vibrating the same amount and the temperature is even throughout the material.

b When does conduction stop and why?

during conduction
transfer of energy

after conduction
equal temperature throughout solid

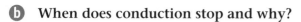

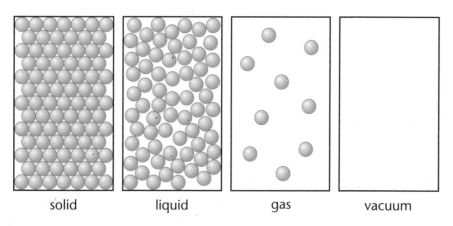

solid   liquid   gas   vacuum

Conduction in non-solids

Look at the diagram on the left. Conduction works best in solids, where the particles are touching and each particle touches many neighbours. In liquids, the particles are touching, but each particle has fewer neighbours to hit against. Conduction is very poor in gases. This is because the particles are far apart in a gas and only hit each other occasionally.

There are some places where there are no particles at all. We call a place with no particles a **vacuum**. You can make a vacuum by pumping all the air out of a container. Much of outer space is a vacuum.

Obviously, conduction does not happen in a vacuum. There are no particles to hit each other, so there can be no transfer of thermal energy by conduction.

Thermal conductors and insulators

Think back to the frozen chicken. How can they reduce the time the chicken takes to heat up?

Some materials conduct thermal energy better than others. The stainless steel draining board, the aluminium foil and other metal objects are good **thermal conductors**. Other materials conduct thermal energy poorly. They are called **thermal insulators**. The cling film, the polystyrene tray and other plastics are thermal insulators.

Non-metal materials that contain gas pockets are very good thermal insulators. Expanded polystyrene, like the tray holding the chicken, is this type of material. So are fluffed-up feathers and woolly jumpers.

c Look at the bottom diagram on page 92.
  (i) Choose a particle in the middle of the solid. How many other particles is it touching?
  (ii) Repeat for the liquid, the gas and the vacuum.
  (iii) Use your answers from *i* and *ii* to explain why solids conduct thermal energy best.

Unwrap it and take it off the polystyrene tray. Plastic is an insulator.

Put it on the metal draining board and cover it in foil. The metal is a good conductor.

d Use you knowledge of conduction to explain why gas pockets make a material a very good insulator.

Questions

1 a Write down the following materials in order, with the best conductor first.

  plastic   air   graphite   aluminium   water

b Explain how you decided on this order.

c Where would a vacuum be on this list? Give your reasons.

2 Use your ideas about particles to explain why kebabs cook more quickly on metal skewers than:

  a with no skewers   **b** on wooden skewers.

3 Look at this cartoon. It represents a solid with one end at a higher temperature than the other. The hands represent the particles and the money represents the energy. Draw or explain the next two frames in the cartoon, showing how energy is transferred during conduction.

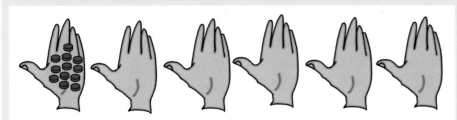

One end of the solid is heated up.

For your notes:

- Thermal energy is transferred from hotter objects to cooler objects.

- **Conduction** is one of the ways in which thermal energy is transferred.

- In conduction, energy is transferred from one particle to the particles touching it, from hotter to colder particles.

- Solids are better conductors than liquids and gases. Some solids are better conductors than others.

16 Convection

Moving air

Think about a hot pie taken out of the oven and left to cool on a wire rack. It is surrounded by air, which is a poor conductor, but it still cools. The air moves, and the particles in the air carry the thermal energy away from the pie. Thermal energy transfer by moving particles is called **convection**. It happens in gases and liquids, because the particles in gases and liquids can move about.

Look at the photo on the right. The candle is heating the air. The hot air then rises, causing the paper spiral to rotate. If you put your hand near the spiral, the air would feel very hot. Thermal energy has been transferred by convection.

a Why does convection happen in gases and liquids but not in solids?

Why does hot air rise?

When you heat air you give the particles more energy. They vibrate more and they move about more. The particles get further apart. The gas expands.

This means that there are fewer particles in the same volume. The diagram below shows the same volume of a hotter gas and a cooler gas. There are fewer particles in the hotter gas.

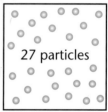 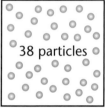

27 particles   38 particles

hotter gas   cooler gas

Fewer particles means less mass. The same volume of hotter gas weighs less than the same volume of cooler gas. It is less dense.

Less dense materials float compared to more dense materials. The hot air rises, like a cork bobbing up through water.

b Look at the hot-air balloon on the left. Why is the air inside the balloon rising compared to the air outside the balloon?

Convection currents

Think about the hot air above the candle. When the hot air rises, something has to take its place. Cooler air moves into the gap. This combination of hot air rising and cooler air falling is called a **convection current**. The diagram on the right shows the convection current around the candle. The convection current mixes the air. Eventually all the air will be heated. When all the air is the same temperature, the convection current will stop.

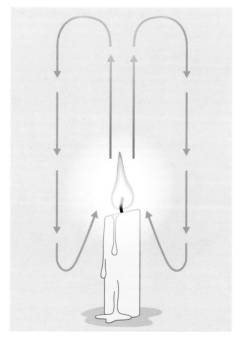

Convection in liquids

Convection also happens in liquids. Look at the photos on the right. They show the same beaker of water being heated. The flame of the Bunsen burner is pointing at the left of the beaker, so only the water there is being heated. Purple dye has been added so that you can see the movement of the water. The photo on the left shows the water after it has been heated for one minute. The photo on the far right was taken after two minutes of heating.

one minute

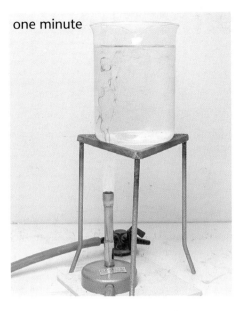

two minutes

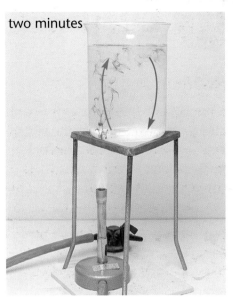

The hot water rises. This can be seen by the purple dye rising. There is a convection current in the water in the beaker. Again, thermal energy is being transferred through the liquid by convection.

c What would have happened to the purple dye by three minutes?

hotter liquid

colder liquid

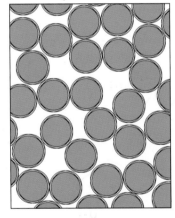

Why do hotter liquids rise?

When you heat a liquid, the particles vibrate more. Each particle takes up more space, so the liquid expands. The diagram on the left shows the same volume of a hotter liquid and a cooler liquid.

d Compare the number of particles in the same volume of the hotter liquid and the cooler liquid.

The hotter liquid has fewer particles in the same volume and is therefore less dense than the cooler liquid. It will rise, in the same way that hot gases rise.

Questions

1 If the air in a hot-air balloon is allowed to cool, the balloon falls. Explain why.

2 During a hot summer's day, the sea is cooler than the land, so the air above the sea is cooler than the air above the land. Explain, using a labelled diagram, how this leads to a cooling breeze blowing in from the sea.

3 Write an explanation of how a heater on one side of a room can heat the air in the whole room.

4 Describe a 'children and money' model for convection. Remember that the children represent the particles and the money represents the energy. (*Hint:* think how particles move in a liquid and in a gas.)

For your notes:

- In **convection** the particles move, transferring the thermal energy.

- Convection happens in gases and liquids, but not in solids.

- A **convection current** happens when one part of a gas or liquid is hotter than another part.

I7 Evaporation, radiation

Learn about:
- Cooling by evaporation
- Thermal transfer by radiation

Suddenly cooler

You cool down very quickly if you stand around while you are wet. This cooling is because water is **evaporating** from our skin.

We also cool down by evaporation when we sweat. Our sweat glands push water out of our pores onto the skin. The water then evaporates from the skin, transferring away thermal energy and cooling our bodies.

Energetic particles

To understand why evaporation causes cooling, you have to think about the particles in the liquid. When you look at the particles in the liquid more closely, you realise that some have more kinetic energy and some have less.

This is shown in the diagram on the right. The darker the red, the more kinetic energy the particle is carrying. The particles with the most energy are the ones that leave the liquid when they reach the surface.

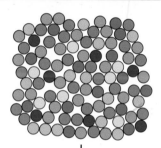

evaporation

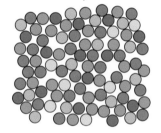

The temperature of the liquid falls. This is because temperature depends on the average energy per particle. If the particles with the most energy leave, there is less energy to be shared between the particles that are left and the temperature of the liquid falls.

You can use the 'children with money' model of energy transfer to think about cooling by evaporation.

There is 200p in the room, split between 20 children. Only three children have the money they need to leave the room. The average money per child is 10p.

There is 152p in the room, split between 17 children. The average money per child is 8.9p.

ⓐ In this model, what represents:
  (i) the energy?   (ii) the particles?
  (iii) the temperature?   (iv) the process of evaporation?

96

Radiation

Thermal energy can be transferred without using particles. This is a good thing, as there is only empty space, a vacuum, between the Sun and us. Thermal energy from the Sun is transferred to the Earth by a process called **radiation**. Radiation is the transfer of thermal energy without particles. During radiation, **infrared radiation** carries thermal energy from a hotter object to a cooler object.

Infrared radiation is like light in many ways. The infrared radiation is produced by a source. It travels away from the source in all directions. All hot objects are a source of infrared radiation. The photo on the right was taken with an infrared camera, which detects infrared radiation rather than light. The people show up brightly, because they are sources of infrared radiation.

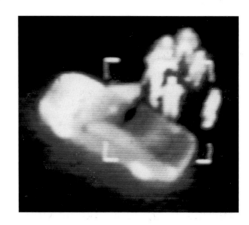

ⓑ The Earth is not heated by conduction or convection.
  (i) Why not?
  (ii) What type of thermal energy transfer heats the Earth?

The Sun produces infrared radiation that heats the Earth. The radiation can travel across the emptiness of space because, like light, it does not need a material to travel through.

Like light, infrared radiation can be reflected by smooth, shiny surfaces. We use this to help stop energy loss from our bodies and our homes. Look at the photo on the left. The shiny survival blanket keeps the athlete warm, reflecting the infrared radiation back towards her body.

ⓒ Ellen is trying to apply the 'children with money' model of energy transfer to radiation. She decides on children with coins standing on one side of a room, throwing coins to children standing on the other side of the room.
  (i) Why doesn't Ellen have any children in the middle of the room?
  (ii) What represents the hot object, the cooler object and the infrared radiation?

Questions

1 Explain carefully how sweating cools the skin using the words below.

  evaporation   **particles**   **energy**   **temperature**

2 Look at this diagram of the cup of tea losing thermal energy to its surroundings. Identify the method of energy transfer shown by each coloured arrow.

3 Use the 'children and money' model to compare conduction, convection and radiation.

For your notes:

- Thermal energy can be transferred from a liquid by **evaporation**. The liquid then cools its surroundings.

- **Radiation** happens when thermal energy is transferred by **infrared radiation**, which is like light.

Think about:
- Changing more than one variable
- Explaining your results

18 Explaining the results

Raising the temperature

Ellen and Sean were investigating how much energy was needed to heat blocks of metal. Each block of metal had two holes. They fitted an electric heater inside one hole and a thermometer inside the other hole.

Ellen and Sean heated each block using the electric heater, and measured the temperature before and after heating using the thermometer. The blocks were insulated to cut down energy losses to the surroundings.

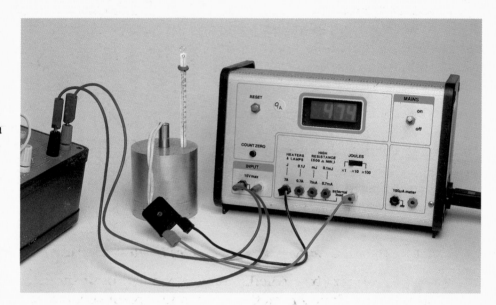

The electric heater was part of a circuit. There was also a joulemeter in the circuit. A joulemeter measures energy. Ellen and Sean measured the energy needed to raise the temperature of each metal block from 25 °C to 65 °C, a temperature rise of 40 °C.

a Analyse Ellen and Sean's results. You may find it helpful to plot a graph.

b Describe the relationship between the energy needed to raise the temperature of the block and the mass of the block.

c Describe the relationship between the energy needed to raise the temperature of the block and the type of metal.

d Use a graph to predict the energy needed to raise the temperature by 40 °C for 200 g of the following metals:
(i) copper
(ii) aluminium
(iii) tin.

e Use a graph to find the mass of each of the following metals that would heat up by 40 °C if you put in 300 kJ of energy:
(i) copper
(ii) aluminium
(iii) tin.

f Bronze is a mixture of 90% copper and 10% tin. Estimate how much energy would be needed to raise the temperature of 1000 g of bronze by 20 °C.

Block	Material	Mass in g	Energy needed in kJ
A	copper	238	368
B	aluminium	214	770
C	copper	164	254
D	tin	284	247
E	tin	341	296
F	copper	372	576
G	aluminium	127	467
H	copper	312	483
I	tin	169	147
J	aluminium	326	1172
K	aluminium	407	1464
L	tin	393	341

Raising and sinking

Justin and Yasmin were taking part in a Science Challenge. The challenge was to make a balloon that would float where it was put, neither rising nor sinking.

Each team was given 12 balloons that had been set up by the organisers. Each balloon was labelled with the type of gas inside, the volume of gas inside and the total mass of the balloon. Some of the balloons sank and some of them rose up towards the ceiling.

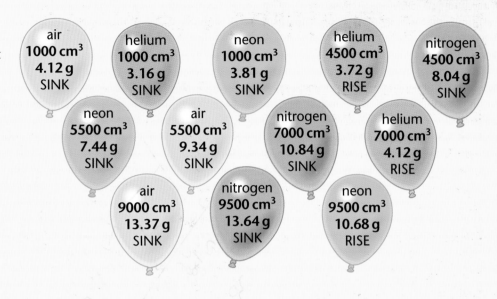

g What are the input and outcome variables for this investigation?

h How does volume affect whether the balloon rises or sinks? Make pairs of balloons from the diagram to explain your answer.

i How does mass affect whether the balloon rises or sinks? Make pairs of balloons from the diagram to explain your answer.

j Choose a balloon from the diagram and suggest ways of making it float rather than rise or sink.

Density

For a balloon to rise it must be less dense that the air around it. If it is less dense it rises up, like hot air in a convection current. During the Science Challenge, the density of air in the room was 0.00115 g/cm^3. Balloons with a density lower than 0.00115 g/cm^3 rose, and balloons with a density higher than 0.00115 g/cm^3 sank.

Density is a variable that depends on two other variables, mass and volume.

$$\text{density} = \frac{\text{mass}}{\text{volume}}$$

Justin and Yasmin decide to make a floating balloon by adding extra mass to the balloon containing 4500 cm^3 of helium.

k What is the density of Justin and Yasmin's chosen balloon without the extra mass?

l What density must the balloon have to float?

m What mass must be added to the balloon to make it float?

Questions

1 All the air-filled balloons in the experiment sank. This would be true no matter how big the balloon. Explain why a balloon filled with air would always sink.

2 Hydrogen has a density of 0.00008 g/cm^3. The balloon itself weighs 3.00 g.

a What volume must 3.00 g occupy before it will float in air with a density of 0.00115 g/cm^3?

b A balloon filled with 2799 cm^3 of hydrogen 'hovers'. Explain why this is different from your answer for **a**.

c Explain why hot-air balloons float using the words below.

mass   volume   density   particles

J1 Magnetic fields

Iron and steel are attracted to magnets. Other materials, including most metals and all non-metals, are not attracted to magnets.

Do you remember?

There is a pulling force between magnets and some metals. The magnet **attracts** the metal.

ⓐ What would happen to the fridge magnets if the fridge door in the photo was made of plastic?

ⓑ Aluminium drinks cans can be recycled, but not steel ones. How can you decide if your drink can is steel or aluminium?

Magnetic magic

Iron and steel are **magnetic materials**. They are attracted to a magnet. The only other metals that are magnetic are nickel and cobalt. Iron oxide is also a magnetic material.

Magnets attract magnetic materials. They can also attract each other. But sometimes the same magnets push away from each other. They **repel** each other.

Look at this photo of a spinning top. It is floating in the air. It looks as if there is nothing holding it up! The magnets in the base are repelling the magnets in the spinning top so it is pushed up into the air. This pushing force balances the weight of the top.

Why do magnets sometimes attract and sometimes repel? Magnets have two different ends, or **poles**. One end is called the **north pole** and the other end is called the **south pole**.

- If you bring the north pole of one magnet towards the south pole of another magnet they attract each other.

- If you bring the north poles of two magnets together they repel each other. Two south poles together also repel.

A magnet will only repel another magnet. It will not repel a magnetic material such as iron.

Opposite poles attract, like poles repel.

ⓒ You have a piece of nickel, a piece of aluminium and a magnet. How could you use a second magnet to tell them apart?

Magnetic fields

Magnets do not only attract or repel at their poles. They pull or push in the space all around them. This space is called a **magnetic field**.

The magnetic field is invisible but we can see where it is by using tiny bits of iron called **iron filings**. The photo on the right shows what happens when we sprinkle these on a piece of paper on top of a **bar magnet**. The iron filings are pulled into lines. These lines show the magnetic field.

The lines are called **magnetic field lines**. They are shown in the diagram on the right.

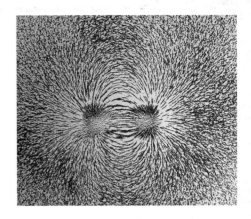

d Why do we need to use iron filings to see magnetic field lines?

How strong?

The magnetic force is strong close to the magnet. It is weaker further away.

Look at the magnetic field lines. Near the magnet they are very close together, showing that the field is strong. Further away they are further apart, showing that the magnetic field is weaker. The magnetic field lines are also closer together at the poles of the magnet than at the sides.

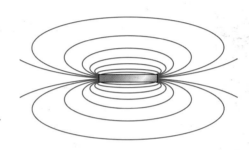

e Where is the magnetic field strongest?

f How can you tell?

A magnet can attract or repel something that is in its magnetic field without touching it. You can see this with the spinning top on page 100.

In all directions

The magnetic field around a magnet reaches out in all directions. Look at this photo of iron filings clustered around a bar magnet. The iron filings line up all around the bar magnet. The magnetic field has height, width and depth. It is three-dimensional.

For your notes:

- Iron, steel, cobalt, nickel and iron oxide are all **magnetic materials**.

- Magnets have a **north pole** and a **south pole**.

- Like poles **repel**, opposite poles **attract**.

- The space around a magnet where its magnetic force works is called a **magnetic field**.

- The magnetic field is stronger closer to the magnet.

Questions

1 Look carefully at the diagram showing the magnetic field lines around a bar magnet.

   a What is the relationship between how close together the lines are and their distance from the magnet?

   b What does this tell you about the magnetic field?

   c Where is the magnetic field strongest, at the poles or at the sides? Explain your answer.

2 Not all magnets are shaped like a bar. Think of a horseshoe-shaped (U-shaped) magnet. Design an experiment to find out the size and shape of the magnetic field around it.

J2 Magnets

Learn about:
- The Earth's magnetism
- The direction of magnetic fields

The Earth's magnetic field

The first scientist who investigated magnetism was William Gilbert. He lived in the reign of Queen Elizabeth I. Years of experiments showed him that if a magnet could move freely, it always pointed north–south.

From these simple observations he made a very large conclusion. He decided that the Earth acted like a giant bar magnet. He published his conclusion in 1600 in a book called *On the Magnet, Magnetic Bodies, and the Great Magnet of the Earth*. The diagram on the right shows the magnetic field of the Earth.

North Pole

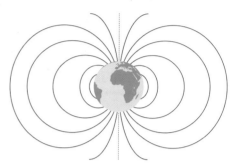

South Pole

We use a **compass** to show where north is. A compass needle is a magnet that can move around freely. The south pole of the magnet will always point towards the Earth's North Pole. This end of the needle is pointed to make it clear that it is pointing north.

A ship's compass from the 1600s.

Did you know?

The Earth acts as a magnet because its core is made of iron and nickel.

(a) Look at the diagram of the Earth's magnetic field. Which way would your compass point if you were standing:
  (i) at the North Pole of the Earth?
  (ii) at the South Pole of the Earth?

You may have noticed two facts that don't seem to fit together:

- The north pole of a magnet points 'north', towards the Earth's North Pole.

- Opposite poles attract. A north pole of a magnet is attracted to a south pole of another magnet.

The idea that came first was that north means at the top of a map, or towards the Earth's North Pole. Then magnets were discovered. The end of the magnet that pointed north was called the north pole of the magnet.

So the 'big magnet' inside the Earth must have its south pole in the north and its north pole in the south. Very confusing! Sometimes science is like that. Blame the person who called the end of a magnet that pointed north a 'north pole'!

Navigating

Having a compass made it much easier for sailors to find their way, or **navigate**. Before compasses, the sailors found their way using the Sun and stars. Bad weather meant they could not see the Sun or the stars, and they often got lost. A compass like the one in the photo works whatever the weather.

Which way does the field go?

A compass always points to the Earth's North Pole, unless you hold it very close to another magnet. Then the pointed end of the compass needle is attracted to the south pole of the magnet.

We can use a compass to show us the directions of the magnetic field lines around a bar magnet. The photo on the right shows compasses close to a bar magnet.

b **Look closely at the photo. Draw a diagram showing clearly where each compass needle is pointing.**

The compasses point along the magnetic field lines. The magnetic field lines start at the north pole of the bar magnet and end up at the south pole of the bar magnet. The arrows show the direction of the field lines in this diagram on the right.

Stopping the field

It is possible to 'stop' a magnetic field if you put the magnet inside a box made of magnetic material. This is called **magnetic shielding**. The box stops the magnetic field, keeping it inside. For magnetic shielding to work, the magnet must not actually touch the box.

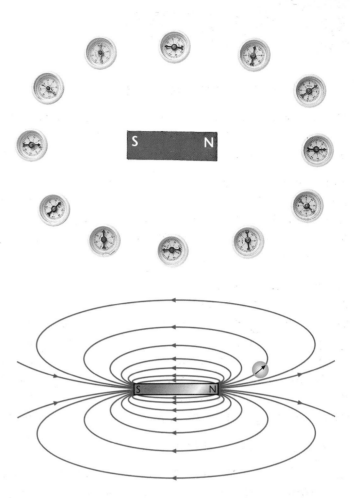

Very strong magnets are used in MRI scanners. These machines, like the one in the photo on the left, are used to look inside the body. The walls of the room are lined with magnetic material. This shielding stops the magnetic field from going beyond the room.

c **Suggest two materials that could be used to make magnetic shielding.**

Questions

1 Explain how you can use compasses to show the direction of the field lines around a magnet.

2 Read these observations about using a compass.

● Compasses always point north.

● Compasses do not work if you are standing at the North Pole or the South Pole of the Earth.

Explain how these observations suggest that the Earth acts as a big magnet.

For your notes:

● The Earth is like a huge magnet. It has a magnetic field.

● A **compass** always points north unless there is a magnet close by.

● A compass close to a magnet points along the magnetic field lines.

● Magnetic field lines run from the north pole of a magnet to its south pole.

● **Magnetic shielding** keeps a magnetic field inside.

J3 Making magnets

Natural magnets

We know that people have been using natural magnets since about 500 BC. **Lodestone** is iron oxide found in rocks. It is a naturally occurring magnet. If you hang a piece of lodestone from a thread, it will always point in the same direction. Lodestone was used as a compass for navigation.

Lodestone.

Making magnets

A paperclip that is touching a magnet picks up more paperclips. It acts as part of the magnet.

If the paperclip was made of iron, it would go back to normal as soon as you take it off the magnet. Iron can only be a **temporary magnet**. It only acts as a magnet when it is in a magnetic field.

Other magnetic materials can be made into **permanent magnets**. They stay as magnets for many, many years. Iron oxide makes excellent permanent magnets.

You can make a magnet using a steel needle. Steel can make a weak permanent magnet. You can use it as a compass needle. The steel needle in the photo on the far right has been magnetised like this. You can see by the compass next to it that it is now pointing north.

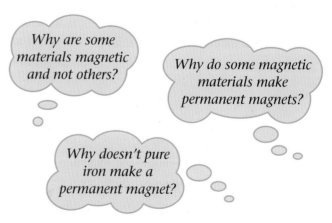

To magnetise the needle you stroke it with one pole of a strong magnet. You have to stroke it several times and always in the same direction. The magnetic field of the strong magnet makes the needle into a magnet.

ⓐ Why do we need to float the needle on water?

Thinking about magnets

If you chop up a piece of lodestone, each of the little pieces acts as a magnet. This is because it is the particles themselves that are magnetic. Each particle has its own tiny magnetic field, with a north pole and south pole.

All the particles in the lodestone have their north poles pointing in the same direction. All the tiny magnetic fields add up, because the magnetic field lines are in the same direction. This gives the piece of lodestone a magnetic field around it and it acts as a magnet.

Why are some materials magnetic and not others?

Why do some magnetic materials make permanent magnets?

Why doesn't pure iron make a permanent magnet?

Kevin: *I'm going to use method 3. It's a complicated set-up but I'll only have to set it up once and then I can increase the current. I can work out the magnetic force from the weight.*

Priya: *I'll use method 2. The tiny washers weigh less than the paperclips, so method 2 should be better than method 1.*

Heidi: *I may as well use method 1. My results will be just 'yes' or 'no'. The cores are too different to compare the strength of the electromagnets.*

d Do you agree with Heidi? Explain your answer.

Results

Heidi's results

Core material	iron	aluminium	nickel	steel	copper	zinc
Paperclips lifted	11	0	4	16	0	0

Priya's results

Number of turns	10	20	30	40	50
Mass of iron at start in g	50.0	50.0	50.0	50.0	50.0
Mass of iron at end in g	47.2	40.6	33.8	20.1	10.4
Iron picked up in g	2.8	9.4	16.2	29.9	39.9

Kevin's results

Current in A	10	20	30	40	50
Mass of nail in g	9.2	9.2	9.2	9.2	9.2
Mass added in g	0	10	20	30	50
Total mass held in g	9.2	19.2	29.2	39.2	59.2
Force in N	0.092	0.192	0.292	0.392	0.592

Analysing the results

e What type of chart or graph could you use to analyse:
  (i)   Heidi's results?
  (ii) Priya's results?
  (iii) Kevin's results?
  Explain your answer.

Heidi refuses to draw a graph. She says that the actual number of paperclips lifted is meaningless, because the cores were all different shapes and masses. She says that all she can tell from her experiment is that magnetic materials make electromagnets when used as a core, and non-magnetic materials do not.

f Do you agree with Heidi's conclusion? Give reasons for your answer.

g What changes would you make to Heidi's method?

h Suggest an extra value for her input variable.

Questions

1 Draw a line graph of Priya's results. Add a line of best fit. Write a conclusion for Priya's experiment.

2 **a** Draw a line graph of Kevin's results. Add a line of best fit.

  b Kevin is happy with his results but Mrs Futter suggests that he should have used smaller masses. Who do you agree with, Kevin or Mrs Futter? Explain your answer.

3 Which was the best method of measuring the outcome variable? Give reasons for your answer.

K1 Seeing the light

Away from the source

Light is energy on the move. At a light source, energy is given out. Look at the photo. Fireworks get their energy from chemical reactions, and some of this energy travels away from the explosion as light. People can see the firework display for many kilometres. This is because light travels away from the source in all directions.

a Where do the following get the energy they give out:
(i) a light bulb? (ii) a candle flame?

Straight through

Light travels in straight lines. We can show this using a camera. The earliest form of camera was a darkened room with a tiny hole to let in light. An upside-down picture or **image** of the scene outside was formed on the wall opposite the hole. We call this type of room a **camera obscura**. The earliest description of the camera obscura was by the Chinese philosopher Mo-Ti, in the fifth century BC.

It is easy to make a small version of a camera obscura, called a **pinhole camera**. This is a box with a pinhole at one end and a screen at the other. The girl in the photo below is using a pinhole camera to make an image of the candle flame on the screen. Just like a full-size camera obscura, the image is upside-down.

The diagram shows what happens. Light from the top and the bottom of the flame travels outwards in all directions. But only the light travelling in the direction of the arrows gets through the pinhole. In the same way, a tiny amount of the light from all parts of the flame goes through the pinhole. This light forms the image on the screen. The image is sharp and upside-down because the light has travelled in straight lines from the source to the screen.

Do you remember?

Light comes from a **source**. Light sources include light bulbs, flames, TV screens, the Sun and the stars. When you block light you get a **shadow**. A shadow is the absence of light.

Did you know?

The Sun gets its energy by turning hydrogen into helium. Some of this energy is given out as light.

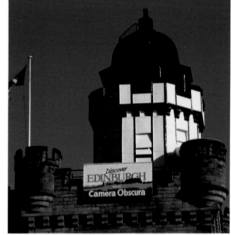

A camera obscura.

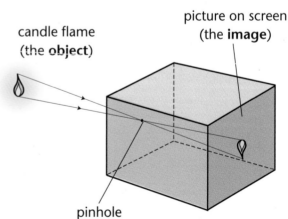

candle flame
(the **object**)

picture on screen
(the **image**)

pinhole

b Would the image be smaller or larger if you used:
   (i)   a shorter pinhole camera?
   (ii) a longer pinhole camera?

c Imagine a pinhole camera with five holes rather than one. What would you see on the screen?

Detecting light

You detect light when it enters your eyes. This is because there are special cells at the back of your eyes that take in the light energy and make an electrical signal that travels to your brain.

Camera film also acts as a light detector. Exposing the film to light makes a chemical change in the film. **Light sensors** also detect light. When light falls on a light sensor a current flows. This means it can be used as a light sensitive switch.

How fast?

Light travels from the source to the detector, usually our eyes. There seems to be no delay between switching on a lamp and seeing the light, so the light must make this journey very quickly. In fact, light travels at 300 000 km/s. That means the distance has to be very great before we would be aware of the time it took. Light takes 8.5 minutes to travel from the Sun to the Earth – a distance of 150 million kilometres.

All light travels this fast. The source does not matter. Candles, TVs and the Sun all produce light travelling at 300 000 km/s. That is 30 million times faster than a man can run, 900 thousand times faster than sound travels and 500 thousand times faster than Concorde flew.

d How long would it take light to travel 1 200 000 km?

Light from the next nearest star takes over four years to reach Earth!

Questions

1 Look at the photo of the lighthouse. To a sailor on a ship the light from the lighthouse appears to be flashing. But the lamp is on all the time and the top of the lighthouse is rotating. Explain why the sailor sees a flashing light.

2 Explain how pinhole cameras suggest that light travels in straight lines.

3 Look at the diagram of the pinhole camera.

   a How many images would you see if you had two pinholes? Draw a diagram to explain your answer.

   b The diagram shows the light stopping at the screen. Look at the photo of the girl using the pinhole camera. Where does the light go after it hits the screen?

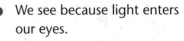

For your notes:

- Light travels away from its **source** in all directions.
- Light travels in straight lines.
- We see because light enters our eyes.
- Light travels very, very quickly.

Light bounces

Why do we see things that are not sources of light? The answer is that light bounces off most surfaces and enters our eyes. This bouncing is called **reflection**. Imagine entering a completely dark room. You see nothing. Then you switch on a lamp and you see everything. The light from the lamp **reflects** off the surfaces and enters your eyes.

Do you remember?

Light sources are **luminous**.
Things that do not make light are **non-luminous**.

a The Moon is not a source of light. Explain why we can see it.

Reflected light

When light reflects off a surface, you can predict where it will go. You can use a thin beam of light, called a **ray**, to investigate reflection.

Look at the diagrams. They show what happens when a ray of light bounces off a flat surface such as a mirror. The arrows show the direction the light is going. The angle between the ray and the mirror is the same for the incoming ray and the reflected ray.

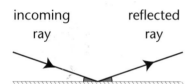

incoming   reflected
ray   ray

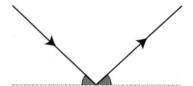

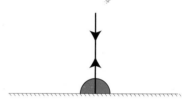

Scientists did not use this angle when they made up their law of reflection. This is because not all surfaces are flat. It is difficult to measure the angle between the rays and a curved or uneven surface. Instead scientists used the angle between the ray and a line called the **normal**.

Look at diagrams **A**, **B** and **C** below. You find the normal by doing an experiment. You aim the ray at the mirror, then change the angle between the ray and the mirror until the ray is reflected back the way it came. The ray was travelling along the normal. For a flat surface, the normal is at 90° to the surface. This is shown in diagram **A**.

Look at diagrams **B** and **C**. The angle between the incoming ray and the normal (the **angle of incidence, i**) is equal to the angle between the reflected ray and the normal (the **angle of reflection, r**).

A
ray coming   ray reflected
in along the   back along
normal   the normal

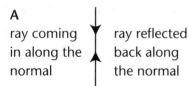

B   normal
incoming   reflected
ray   ray
          i *r*

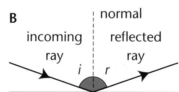

C   normal

          i *r*

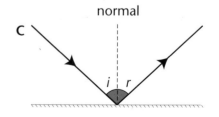

b What is a ray?

c How do you find the normal?

Scattering

Most surfaces are not perfectly flat and smooth. Even the paper this book is made of has many tiny bumps on its surface. The photo shows paper under a microscope. When light hits the paper, it is scattered in all directions. The diagram shows this scattering.

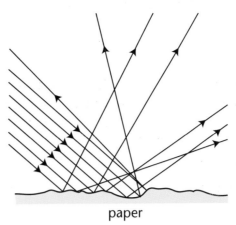

paper

Reflectors on bicycles have uneven surfaces so that they scatter the light in many directions. They are also very reflective, so they reflect a lot of light. This makes the cyclist very visible.

Mirrors

A mirror has a perfectly smooth surface. Look at the diagram of rays of light hitting a mirror. The light rays come in from the same direction and the rays are all reflected in the same direction. This is why you can see an image of your face when you look in a mirror but not when you look at a piece of paper.

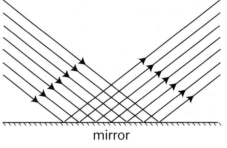

mirror

d **Explain why a dirty knife will not act as a mirror, but a clean knife will.**

The image in a mirror is the wrong way around, or **inverted**. Left is right and right is left. This is why ambulances have the word AMBULANCE on the front, so that drivers can read it in their mirrors.

Using mirrors

You can use mirrors to see around corners, or over a wall. Look at the diagram. This periscope uses two mirrors so you can see what is happening above you.

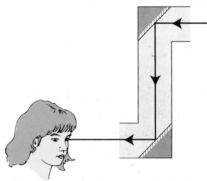

A periscope.

Questions

1 a A puddle can act as a mirror. Explain why, using a diagram.
   b The image is broken up when a pebble drops into the puddle. Why is the image broken up?

2 Make a careful, large copy of the diagram of the periscope (using squared paper will make it easier).

   a Mark in the normal at each mirror.
   b Looking at each mirror in turn, which two angles are equal? Show this on your diagram.
   c Explain in your own words how the periscope could help you see over a wall.

3 In a laser light show, you see beams of coloured light travelling up into the sky. Suggest how you see the beam of light when it is travelling from the laser up into the sky. (*Hint:* smoke or mist makes the beams easier to see.)

For your notes:

- Light is **reflected** from many surfaces.

- During reflection, the **angle of incidence** equals the **angle of reflection**.

- Most surfaces **scatter** light when they reflect it.

- Mirrors do not scatter light because they are smooth.

- The image in a mirror is the wrong way round, or **inverted**.

Absorption or transmission?

Light is not always reflected when it hits a surface. When light hits **opaque** materials like wood or bricks, it is 'soaked up'. This is called **absorption**. The light energy is transferred into heat energy.

When light hits **transparent** materials like glass and water, it travels through them. We say the light is **transmitted**. Sometimes part of the light is absorbed and part is transmitted. Materials that both absorb and transmit light are called **translucent**. Thin paper is translucent.

A trick of the light

You often see strange effects when light passes through a transparent material like water or glass. Swimming pools look shallower than they really are. A pencil in a bowl of water looks bent. Text looks closer and larger when looked at through a glass block.

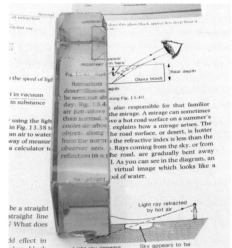

Air to glass

These strange effects happen because the light bends when it goes from one material to another. We call this bending **refraction**. We can investigate refraction using a block of glass.

Look at diagram A. It shows what happens when a ray of light enters glass at 90°, along the normal. The light does not bend. Look at diagrams **B** and **C**. When the ray of light hits the glass at an angle, it bends. It bends towards the normal.

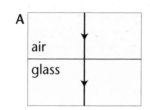

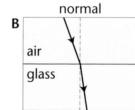

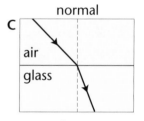

Why does light bend?

The light refracts or bends because it travels at different speeds in different materials. Light travels faster in air than in water or glass. When the light hits the boundary at 90°, all the light slows down at the same time, so the ray stays straight. If the light hits the boundary at an angle, some of the light slows down first, making a bend.

One way of thinking about this is to imagine a truck driving into mud. If the truck hits the mud straight on (along the normal), then both sides of the truck slow down at the same time and the truck goes in a straight line. If the truck hits the mud at an angle, then one wheel slows down before the other, and so the truck swings around.

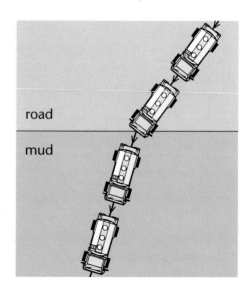

a Light refracts when it goes from air to glass at an angle. Does it bend towards the normal or away from the normal?

b Does light refract when it goes from air to glass along the normal?

Water to air

Refraction also happens when light travels from glass to air, or from water to air. Look at the diagram. It shows a coin at the bottom of a beaker of water. Light is reflected from the coin. The arrow shows a ray of this light. As the ray passes from water to air, it bends or refracts. This time it bends away from the normal, because light travels faster in air than in water.

Your brain thinks that light travels in straight lines. You see the coin higher up, where the light appeared to come from.

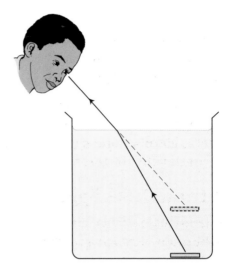

Archer fish hunt by shooting water at insects flying above the water. The archer fish have to learn to allow for refraction.

c Draw a ray diagram of an archer fish looking at an insect. (*Hint:* use the diagram of the coin in the beaker to help you, but remember that the light travels from the insect to the fish's eye.)

d Will the archer fish have to shoot high or shoot low to hit the insect?

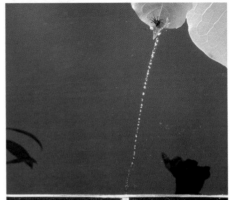

Refraction is not a problem for an archer fish.

Questions

1 Refraction does not always happen when light goes from one transparent material to another. What two things must happen for refraction to take place?

2 Light travels at different speeds in air, water and perspex. It travels fastest in air, slower in water and slowest in perspex.

   a When light travels at an angle from air to water, the light slows down and refracts towards the normal. Predict what happens when light travels at an angle from:

   (i) water to air   **(ii)** air to perspex   **(iii)** water to perspex.

   b Of the experiments described above, which will make the light bend most?

For your notes:

● Light is **absorbed** and/or **transmitted** by some materials. These materials can be **transparent**, **translucent** or **opaque**.

● When light goes from one transparent material to another, it may **refract** (bend).

● The light must enter the new material at an angle for refraction to happen.

● Light has to travel at different speeds in the different transparent materials for this to happen.

Splitting white light

When you pass sunlight through a prism, the white light splits up into many colours. We call the colours produced a **spectrum**. The splitting up is called **dispersion**. White light is made up of many colours.

The colours are always in the same order: red, orange, yellow, green, blue, indigo and violet. The light has refracted twice, once when the light entered the glass and again when the light re-entered the air. The violet light bends most and the red light bends least. The difference in refraction spreads the colours.

a Which colour refracts the most?

b Which colour refracts the least?

A rainbow is made because water droplets in the air act like tiny prisms. Sunlight is refracted as it enters and exits the water. Different colours are refracted by different amounts, so the colours spread out, making a rainbow.

You can mix the different colours together again using a second prism. Mixing the colours gives you white light. You can read more about this on page 118.

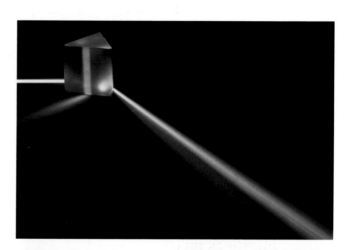

Coloured filters

Look at the photo of a stained glass window. White light falls on the window, but coloured light comes through the window and enters our eyes. Some colours in the light are absorbed by the coloured glass, and some colours are transmitted.

The red glass transmits red light and absorbs the other colours. In the same way, the green glass transmits green light and absorbs the other colours.

Did you know?

You can remember the colours of the spectrum like this:

Richard **O**f **Y**ork **G**ave **B**attle **I**n **V**ain.

c Look at the diagram. What would you see if the light shone onto the window was:

(i) red light?   (ii) blue light?   (iii) green light?

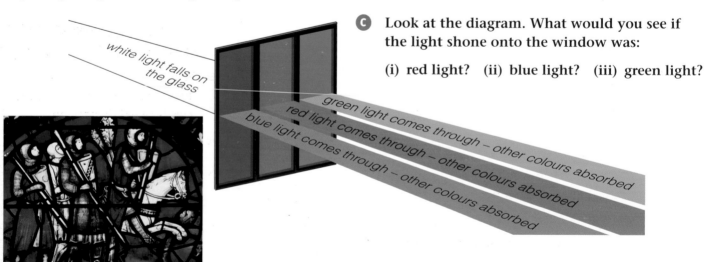

white light falls on the glass

green light comes through – other colours absorbed

red light comes through – other colours absorbed

blue light comes through – other colours absorbed

Seeing by reflection

A red surface reflects red light. The red light enters your eyes and you see red. If you shine white light onto a red surface, the red light is reflected and the other colours are absorbed.

Photographers develop their film in a darkroom. They use a special red light to see what they are doing. The red light does not affect the film.

Things look odd when the red light is on. The photographer is wearing a white shirt, but it looks red. She is wearing green trousers, but they look black.

d (i) Why does the white shirt look red?
(ii) Why do the trousers look black?
(iii) What happens to the red light when it hits the trousers?

Plants reject green

When we think of plants we think 'green', but plants do not use green light to make their food. Green leaves reflect green light and absorb the other colours, using the energy to make their food. The reflected green light enters our eyes, so we see green leaves.

Questions

1 Conor is investigating light using coloured filters. He does the experiments shown in the diagram. What colours of light (if any) will he see at **A**, **B**, **C** and **D**?

Experiment 1   Experiment 2   Experiment 3

white light → A?   green light → B?   white light → C? → D?

2 Ellen goes to a club. The club is having a 70s night with a disco. The lights flash green, blue and red. Ellen is wearing a white shirt, a red skirt and black boots.

Predict what Ellen's shirt, skirt and boots will look like when:

a the red light is on
b the blue light is on
c the green light is on.

Give an explanation for each of your answers.

3 Plants need light to make food. What would happen if you tried to grow plants under green lights? Explain your answer using the words absorb, transmit and reflect.

For your notes:

● White light can be split into a **spectrum** of different colours. This is called **dispersion**.

● A red, green or blue filter allows only one colour of light through and absorbs the others.

● A coloured object reflects the colour we see and absorbs the other colours.

Combinations

Joe's class has been collecting copper coins for a charity. Joe is counting the money and putting it into bags. There are three possible **combinations**:

- only 1p coins in a bag
- only 2p coins in a bag
- a bag of mixed 1p and 2p coins.

The coins can be combined in three different ways, to make three different combinations. You can read more about combinations in the blue box on the opposite page.

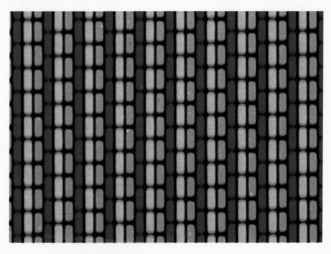

Primary colours of light

The photo shows a television screen magnified many times. You can see that the screen is made up of dots. There are three colours of dot: red, green and blue.

Red, green and blue are the three **primary colours** of light. All colours of light can be made from combining these colours.

a How many different combinations can you make out of red, green and blue?

Secondary colours

Secondary colours of light are made by mixing red, green and blue light. Look at the colour chart on the right. It shows what happens when you combine red, green and blue light. For example, colour **E** is made by combining **A** with **C**.

b Copy this table and use the colour chart to fill in the gaps.

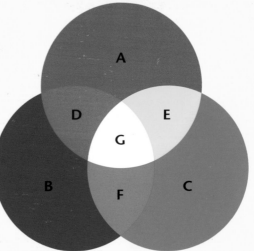

Position	Colour	Which combination of red, blue, green
A	red	
B	blue	
C	green	
D	magenta	
E	yellow	
F	cyan	
G	white	

Lighting the show

Zahir wanted to help with the lighting for school productions. Miss Lawrence said that he had to learn about colour. She set him a test.

c Do the 'Lighting test'. It is in the green box on page 119.

d Mark your answers (see the bottom of page 119).

Zahir scored 7 out of 10. He could not do questions 8, 9 and 10.

e Write down what you would say to explain questions 8, 9 and 10 to Zahir. Draw any diagrams that you would use.

An example of combinations

Jane has bought some sweets. She has toffees, cream eggs, boiled sweets and humbugs. How many different combinations can she make?

Jane can have each type of decoration on its own (■ toffees ■ cream eggs ■ boiled sweets ■ humbugs).

Jane can have two types of sweet together (■ toffees and cream eggs ■ toffees and boiled sweets ■ toffees and humbugs ■ cream eggs and boiled sweets ■ cream eggs and humbugs ■ boiled sweets and humbugs).

Jane can have three types of sweet together (■ toffees, cream eggs and boiled sweets ■ toffees, boiled sweets and humbugs ■ cream eggs, boiled sweets and humbugs ■ toffees, cream eggs and humbugs).

Jane can have all four types of decoration together (■ toffees, cream eggs, boiled sweets and humbugs).

So there are 15 different types of combination.

Lighting test

Imagine you have only three spotlights: red, green and blue.

What combinations of red, green and blue do you use to make the following colours?

1 yellow   **2** cyan   **3** magenta   **4** white?

You can change the colour of a spotlight by putting in different filters.

What combination of primary colours gets through each of the following filters?

5 red filter   **6** yellow filter   **7** cyan filter?

You can put more than one filter in a white spotlight. What colour of light would you get with these combinations of filters:

8 yellow and cyan   **9** magenta and cyan

10 magenta and yellow?

Seeing colours

We see colour because we have light receptors called cones in the backs of our eyes. We see three primary colours because we have three types of cone. One type is sensitive to red light, one to green light, and one to blue light. People who are colour-blind have a fault in one or more of these types of cone.

Some fish have four types of cone.

f How many primary colours will these fish see?

g Use your understanding of combinations to work out how many colours (both primary and secondary) the fish will see.

Questions

1 Bernie is putting up fairy lights to celebrate the New Year. He has five different types of bulb that he can use:

- red flashing
- red
- green flashing
- green
- white.

What different combinations of bulbs could Bernie put in his string of fairy lights? List them all.

Answers to Lighting test
1 green and red
2 green and blue
3 red and blue
4 red, green and blue
5 red
6 red and green
7 blue and green
8 green
9 blue
10 red

L1 Good vibrations

Making music

Jon and Letticia investigate sound using a kettledrum.

ⓐ **What part of the drum vibrates to make the sound?**

Jon suggests that the pitch of the sound is to do with how fast the drumskin vibrates. Letticia suggests loudness is to do with how big the vibrations are.

Testing their ideas

Their teacher says that they can test their ideas using a **microphone** and a **cathode ray oscilloscope**, or **CRO**. The microphone turns the sound into an electrical signal. The CRO then shows the electrical signal on the screen.

Pitch

The pupils then use tuning forks to make sounds with different pitches. They observe the electrical signals on the screen of the CRO.

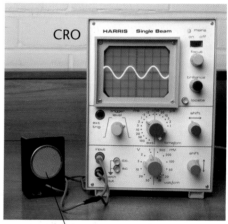

CRO

microphone

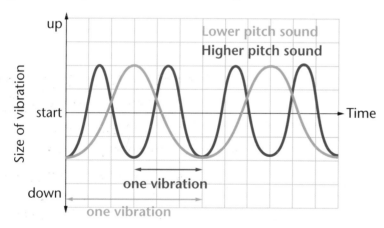

Jon makes a low-pitched sound using a tuning fork and observes the pattern on the CRO screen.

Letticia then makes a high-pitched sound with a tuning fork and observes the CRO pattern for her sound.

The size of the vibrations is the same. The number of vibrations in the time shown is different. The lower-pitched sound shows two vibrations. The higher-pitched sound shows four vibrations. The higher-pitched sound has more vibrations in the same time. We say it has a higher **frequency**.

Higher-pitched sounds have a higher frequency, or more vibrations per second. Lower-pitched sounds have a lower frequency, or fewer vibrations per second. Frequency is measured in **hertz** or **kilohertz**. A sound with 256 000 vibrations per second has a frequency of 256 kilohertz. This sound is middle C on a piano.

ⓑ **In what way are the two sounds shown in the graph the same? Why was this important for the experiment?**

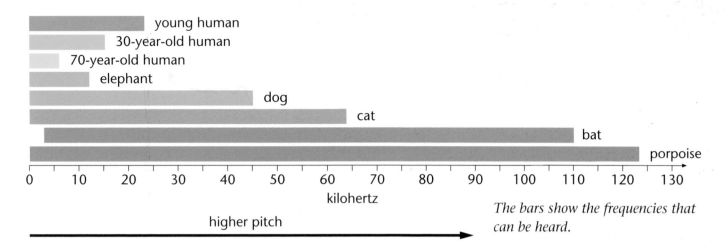

kilohertz

higher pitch →

The bars show the frequencies that can be heard.

The line graph shows hearing loss in humans from age 10 to 80. Our hearing is best when we are teenagers. Gradually we lose the ability to hear high-pitched sounds.

d Describe what is:
   (i) similar in the hearing of all age groups
   (ii) different in the hearing of all age groups.

e At what age does the hearing of middle-pitched sounds get much worse?

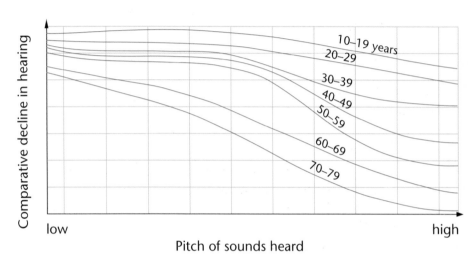

Hearing loss from age 10 to 80.

Questions

1 The table shows the lowest frequency sounds that can be heard by the six mammals in the bar chart.

a What unit is the frequency measured in:
   (i) the bar chart?
   (ii) the table?

b Presenting the data in a table and in a graph tells us something about the way animals hear low frequency sound.

Mammal	Lowest frequency that can be heard, in hertz
human	64
elephant	16
dog	67
cat	45
bat	2000
porpoise	75

Which method is better suited for this purpose? Explain.

2 Imagine you are designing a mobile telephone system. The wider the range of frequency the telephone transmits, the more expensive it will be to build. What range of frequencies would you design for?

For your notes:

● The **eardrum** vibrates when sound enters the ear.

● Damage to the eardrum can cause **hearing impairment**.

● Humans can't hear some sounds of very high or very low frequency.

● As humans get older, they lose their ability to hear high-frequency sounds.

125

Living with noise

Look at the photos. They show people who work where the sound is very loud. We call the sound from the chainsaw '**noise**' because it is sound no-one wants. We call the sound in the club 'music'. People enjoy listening to music. Whether it is music or noise, loud sounds can damage your hearing.

We measure loudness with a sound meter. Loudness is measured in **decibels**. A very quiet whisper may be 1 decibel. A loud sound like a vacuum cleaner is 70 decibels. A jet plane overhead is about 100 decibels.

Bang!

Sounds of over 120 decibels, like those made by an explosion, can break the eardrum. Too much energy is passed to the eardrum so it stretches and breaks. We call the vibrations that travel away from an explosion a **shock wave**.

The damage caused by a shock wave is even worse if you are underwater. The vibrations travel more quickly underwater and carry more energy.

(a) **Explain these differences between shock waves travelling through air and water:**
   (i) **the speed of the shock wave**
   (ii) **the energy carried by the shock wave**

We measure loudness with a sound meter.

Too loud

Very loud sounds of 90–110 decibels also cause hearing impairment. After going to a nightclub, the clubbers may not hear very well. Normal sounds seem quiet. They may ask people to speak up. Their ears have become numb. The numbness wears off quite quickly, and by next morning their ears will work again.

If this happened again and again the numbness may become permanent. People who work with very loud noises every day are in danger of damaging their hearing. That is why the man with the chainsaw is wearing ear-protectors. It is why some rock stars and DJs lose their hearing as they get older.

There are very strict regulations about loud sound. Employers have to cut down the noise as much as they can. They also have to provide ear-protectors.

Loud sounds may damage the inner ear, or the nerves that carry the signals from the inner ear to the brain.

Cutting noise

Noise can cause irritation and stress even if it is not loud enough to damage people's hearing. **Sound insulation** lets people make the noise they want without annoying others. It 'soaks up' the sound vibrations.

When sound hits a surface it is reflected, transmitted or absorbed, in the same way as light is reflected, transmitted or absorbed. We sometimes hear the reflected sound as an **echo**. Usually the reflected sound is scattered, increasing the level of noise.

We can reduce noise by preventing the sound being reflected or transmitted, and increasing the amount of sound being absorbed. Some materials absorb energy better than others. Carpets and curtains absorb more sound energy than bare walls and floors. Bare floors and walls reflect the sound that hits them.

Scientists have developed materials that are particularly good at absorbing sound. Rubber and foam absorb a lot of sound energy. They also work on surfaces with special shapes, which do not reflect the sound back into the room.

b Music in some clubs is played at about 110 decibels. Why is this more dangerous to the people who work there than to the clubbers?

All the sound in this room is absorbed by the foam pyramids covering the floor, walls and ceiling.

Questions

1 Describe and explain the three effects that loud sound can have on hearing.

2 Match the sound with the correct number of decibels.

   loud thunderclap   busy street   chatting   whispering

   60   3   70   110

3 Mary's parents are thinking about confiscating her drum kit. What could Mary do to stop the noise leaving her room? Use the words **reflect, transmit, absorb** and **energy** in your answer.

For your notes:

- Very, very loud sounds can break the eardrum and stop the ear working.

- Loud sounds can cause hearing impairment. If they go on a long time, this can be permanent.

- **Sound insulation** absorbs the vibrations, stopping the sound.

127

L5 Detect it

Think about:
- Collecting information

Senses

There are five senses: sight, hearing, touch, smell and taste. We use our senses to collect information about our surroundings. Compared with other animals, humans' best sense is probably sight. We see in colour. We are good at seeing movement. We can judge distance.

Scientists need to collect information. Although their senses are good, they are sometimes not good enough. They build special equipment to extend their senses.

a **Use your general knowledge to answer this question. What scientific equipment do scientists use to detect things that are:**
(i) too small to see?
(ii) too far away to see?
(iii) inside the human body?

Seeing the sound

There are many sounds that humans cannot hear.

You may have to look up information on other pages of this book to answer **b–e**.

b **Which of these sounds are too quiet for humans to hear?**

100 decibels 10 decibels 1 decibel 0.1 decibel 0.01 decibel

c **Which of these sounds are too high frequency for humans to hear?**

1 kilohertz 10 kilohertz 100 kilohertz 1000 kilohertz

d **Which of these sounds are too low frequency for humans to hear?**

1 hertz 10 hertz 100 hertz 1 kilohertz 10 kilohertz

We can use a microphone and a CRO to see sounds we can't hear.

e **The graph shows the pattern for a sound on a CRO.**
(i) For this sound, how many vibrations are there in 0.0001 seconds?
(ii) How many vibrations would there be in 0.001 seconds?
(iii) How many vibrations would there be in 0.01 seconds?
(iv) How many vibrations would there be in 1 second?
(v) What is the frequency of this sound?
(vi) Could you hear this sound?

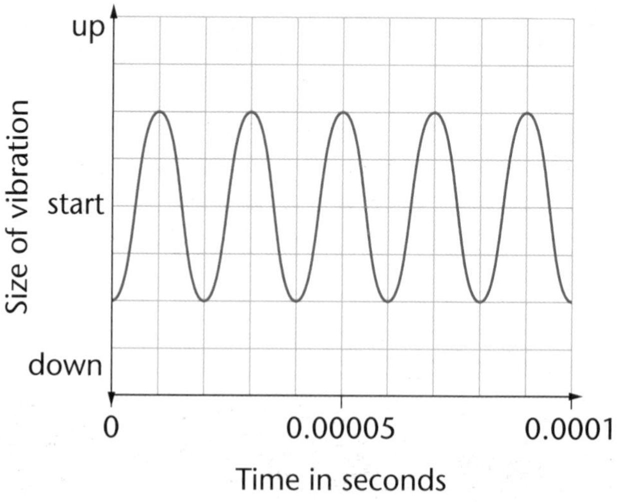

Seeing the heat

We can detect heat energy, but only when the hot object is very close to our skin, or very, very hot (like the Sun).

We use an infrared camera to see heat energy at a distance. The camera changes the heat energy into light for us to see.

f **Suggest a use for infrared cameras.**

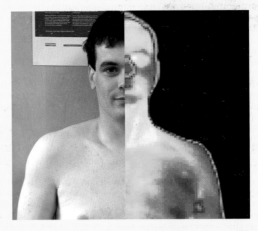

Other instruments

Scientists have invented other scientific instruments so that we can detect more types of information.

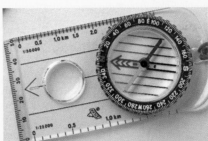

The compass detects the magnetic field by pointing towards north. We cannot detect a magnetic field without it, but pigeons can!

This chromatography machine is more sensitive than a dog's nose. It detects tiny amounts of chemicals in the air. It shows the information as a chart that we can see.

This geiger counter detects radiation from radioactive substances. We can't detect this radiation, but it is very dangerous. The instrument gives the information as 'clicks' that we can hear, and on a dial which we can see.

g **Why do you think so many scientific instruments change the information into something we can see?**

Questions

1 a Make a list of all the scientific instruments you can find on these pages, or that you have mentioned when you answered a question. There are seven.

b Add 'ammeter', 'pH meter', 'voltmeter' and 'thermometer' to your list for **a**. Put them in a table like this, and fill in the other columns.

Instrument	Detects	Produces information as ...
geiger counter	radiation	sound and light

2 Dogs see in dimmer lights than humans. They see movement better than humans. Everything at a distance is blurred to a dog. They see yellow and blue but not red and green: red and green probably look grey to a dog. Most dogs' eyes are about 30 cm from the floor. Draw a picture of what the world might look like to a dog.

3 Our scientific instruments take in information about the environment and change it into light energy, so we can see it. Imagine if dogs became scientists.

a Do you think their scientific instruments would give out information the dogs could see?

b Choose a scientific instrument and design a doggy version.

Glossary

absolute zero The lowest temperature possible, when all the energy has been taken away from an object.

absorbed/absorption The process by which digested food passes through the lining of the small intestine into the blood.

When light or sound is soaked up by a surface, it is absorbed.

active immunity The body produces antibodies against microbes in a vaccination.

adaptations Having features that help a living thing to survive in a particular place.

aerobic respiration The process by which plants and animals use oxygen from the air to break down their food to release the chemical energy from it.

agar plate A plate containing agar jelly used by scientists to grow bacteria.

air resistance The friction a moving object makes with air.

alchemists Early scientists who tried to turn substances into gold.

alveoli Tiny, thin-walled air sacs in the lungs.

amplitude A measure of the size of a vibration. The distance on a graph between the midpoint and the furthest point of the vibration.

amylase An enzyme in saliva that helps to break down starch.

anaerobic respiration The process by which plants and animals break down their food to release the chemical energy from it, without using oxygen from the air.

angle of incidence The angle between the incoming ray and the normal.

angle of reflection The angle between the reflected ray and the normal.

anorexia nervosa An eating disorder which causes a person to eat too little.

antibiotic A medicine that kills bacteria but has no effect on viruses.

antibodies Special chemicals, produced by white blood cells, which attach themselves to the outside of microbes and kill them.

antiseptics Chemicals that kill bacteria.

antivirals Drugs used to treat diseases caused by viruses.

anus The opening through which undigested food passes out of the body.

arteries Blood vessels in which blood flows away from the heart.

atom The simplest type of particle.

attracts Pulling force between magnets and some metals.

bacteria Unicellular microbes with a cell wall but no nucleus.

balanced diet A diet that has the right amounts of all the nutrients.

balanced forces Two forces of the same size pulling in opposite directions.

bar magnet A rectangular magnet.

basalt A type of igneous rock with small crystals.

biological washing powder Washing powder that contains enzymes to break down stains caused by proteins.

biological weathering Breaking down rock by the action of plants or animals.

blood vessels Tubes in which blood flows all around the body.

boiling point The temperature at which a liquid element becomes a gas.

breathe The process of getting air in and out of our lungs.

broad-spectrum antibiotics Antibiotics that kill a wide range of bacteria.

°C The short way to write degrees Celsius.

calcium A mineral found in milk and cheese, and it keeps your teeth and bones healthy. The mineral calcium is actually calcium compounds.

camera obscura The earliest form of camera – a darkened room with a tiny hole to let in light. An upside-down image of the scene outside was formed on the wall opposite the hole.

capillaries Very small blood vessels.

carbohydrates Nutrients found in foods such as bread, which give you energy.

carbon dioxide A waste product of respiration. A gas that is produced when carbon burns and joins with oxygen.

cathode ray oscilloscope (CRO) Machine that shows the pattern of sound on a screen.

cells Tiny building blocks that make up all living things.

cellulase The enzyme that breaks down cellulose.

cellulose Molecules in plant cell walls.

cell membrane A thin layer that surrounds the cell and controls the movement of substances in and out of the cell.

cell wall A tough box-like wall around plant cells.

Celsius A temperature scale in which 0°C represents freezing water and 100°C represents boiling water.

cementation In the gaps between compacted grains in sedimentary layers, chemicals in the water crystallise and 'glue' the grains together.

changes of state Changing from a solid to a liquid or a liquid to a gas and back again – melting, freezing, boiling, condensing.

chemical changes Irreversible changes in which new substances are formed.

chemical energy Energy stored in a material, which will be given out in a chemical reaction.

chemical process A change that makes a new substance.

chemical properties The way a material changes during chemical reactions.

chemical weathering The breaking up of rocks by chemicals in the environment. The substances in the rocks are changed into new substances.

chlorophyll A green substance that is needed for photosynthesis.

chloroplasts The parts of a plant cell that carry out photosynthesis.

circulatory system An organ system that transports substances around the body.

closed question A question that has only two possible answers, often 'yes' or 'no'.

combinations Ways in which objects are put together.

combustion The chemical reaction that happens when something burns.

compaction Grains in sedimentary layers pressed tightly together by the enormous weight of layers of sediment deposited later.

compass Instrument used for navigation, with a magnetic needle which points to the Earth's North Pole unless there is a magnet close by.

competition The struggle between organisms in a habitat for scarce resources, eg food, water, space.

composition The parts something is made from.

compound A substance with more than one type of atom joined together.

conduction Thermal energy is passed from particle to particle in a solid.

cones In plants – structures that contain the seeds in conifers.

conifers Plants that reproduce from seeds in cones and have thin, needle-like leaves.

constipated Difficulty in emptying the bowels.

consumer An animal, that eats (consumes) plants or other animals.

contract A material getting smaller.

control A second experiment where the variable being investigated in the first experiment is held constant.

convection Particles move, transferring thermal energy from particle to particle.

convection current A circular movement of hot gas (or liquid) rising and cool gas (or liquid) falling.

core Magnetic material placed inside a coil of wire (with an electric current running through it), to make the magnetic field stronger.

correlation A link between two or more things.

CRO Cathode ray oscilloscope.

crystalline A substance that contains crystals is crystalline.

crystals Groups of molecules with a symmetrical structure.

cuticle A waterproof layer on the surface of some leaves.

cytoplasm A jelly-like substance found inside cells.

datalogger Equipment with sensors to measure different variables over time, eg temperature.

decibels A measurement for the loudness of sound.

decomposers An organism that feeds on the dead bodies of plants and animals.

degrees Celsius A temperature scale in which 0 °C represents freezing water and 100 °C represents boiling water.

delta New land formed by deposition at the mouth of a river.

dense A dense material has a lot of particles in a small volume.

density How heavy a material is for its size (density = $\frac{\text{mass}}{\text{volume}}$).

deposition Small pieces of rock settling at the bottom of a river or the sea.

diffusion Gas or liquid particles spreading out as their particles move and mix.

digestion Process by which food is broken down into smaller molecules.

digestive juices Juices in the digestive system that contain enzymes which help to break down different nutrients.

digestive system The organ system that breaks down your food into smaller molecules and absorbs them.

dispersion The splitting of white light into colours.

double-blind trial An experiment to discover reactions, eg to a drug, where neither the doctors nor the patients know which is the real medicine and which is the placebo.

eardrum Part of the ear that vibrates when sound reaches it.

ecosystem An area such as a forest or a pond, including all the living things in it and also its soil, air and climate.

egestion Passing undigested food out of the body.

electrical energy Energy carried by electricity.

electromagnet A magnet that can be switched on and off using electricity.

element A substance that cannot be broken down into anything simpler.

energy Energy makes things work. When anything happens, energy is transferred.

environment The surroundings in a habitat.

enzymes Proteins that speed up the breakdown of food in digestion.

epidemic A disease that attacks many people at the same time in a community.

erosion Loose pieces of rock are broken down while being transported.

estuary Where a river flows into the sea.

evaporating/evaporation The change of a liquid into a gas using thermal energy transferred from the liquid's surroundings.

expand A material getting bigger.

extrusive Igneous rock with small crystals, formed when lava cools quickly on the surface of the Earth, such as basalt.

faeces The undigested food that is egested from the body.

fats Nutrients found in foods, such as butter, that give you energy and insulate your body.

fermentation Anaerobic respiration carried out by yeast, which produces ethanol.

ferns Plants that reproduce from spores and have leaves called fronds.

fibre Bulky material found in cereals, fruits and vegetables that helps to keep food moving through your gut.

fixed composition Always made from the same atoms, present in the same ratio.

flowering plants Plants that reproduce from seeds made in flowers and have various shapes of leaf.

food chain A diagram that shows how the organisms in an ecosystem feed on each other.

food web Lots of food chains linked together to show the feeding relationships in an ecosystem.

force An action on something that causes it to move, change direction or change shape.

forces of attraction Pulling forces between particles that hold them together.

formula Symbols used to represent atoms in a compound.

fossils The remains of animals or plants that have been buried deep underground for millions of years and preserved.

freeze-thaw weathering Water inside a crack freezes and expands, exerting a force on the rock. The ice then thaws. This process is repeated many times until the rock eventually breaks apart.

frequency The number of vibrations per second measured in hertz. Higher-pitched sounds have a higher frequency.

friction The force that is made when things rub together.

fronds The large tough leaves of ferns.

fungi (_singular_ **fungus)** Living things that feed on rotting material, for example toadstools.

gas exchange In alveoli, the movement of oxygen into the blood and carbon dioxide out of the blood.

geologists Scientists who study the Earth and rocks.

glacier A slow-moving river of ice that can erode rocks by scraping across the top of them.

glucose A small molecule formed by breaking down carbohydrates, or made by plants in photosynthesis.

goitre A disease caused by not eating enough iodine in the diet, in which the thyroid gland in the neck becomes very large.

grains Tiny pieces of material, such as sand.

granite A type of igneous rock with big crystals.

gravitational energy Energy stored because something is lifted up.

groups The eight vertical columns in the periodic table.

gullet The part of the gut that links the mouth and stomach, also called the oesophagus.

gut The long tube in your body down which food passes between the mouth and the anus, and where digestion and absorption take place.

habitat A place where an organism lives, that provides all the things the organism needs to carry out the life processes.

hearing impairment Damaged hearing.

hertz A measurement for pitch.

humus Dead animal and plant material found in soil. It provides plants with nutrients.

hydrogen peroxide A compound containing two hydrogen and two oxygen atoms.

hydrogencarbonate indicator A very sensitive indicator for the level of carbon dioxide.

hyphae Long microscopic threads in some fungi, such as mould.

hypothesis A possible explanation for why something happens.

igneous rock Rock that is formed from molten lava or magma that has cooled and solidified.

image An object seen indirectly on a screen or using a mirror or lens.

immune Protected against infection.

immune system The body's defences against infection.

immunised Protected against a disease caused by microbes.

infections Diseases.

infrared radiation Carries thermal energy from a hotter object to a cooler object.

inoculation A treatment with microbes to provide immunity against a disease.

interdependent Dependent on each other.

interdependence Organisms in the same food web all depend on each other.

interlocking Crystals in rocks which fit together, with no gaps between them, are interlocking.

intrusive Igneous rock with large crystals, formed when magma cools slowly underground, such as granite.

inverted Upside down.

iodine A mineral found in fish, used by the thyroid gland to make a hormone that helps you grow.

iron A mineral found in liver and eggs, and it is used in your body to make blood.

iron filings Tiny shavings of iron used to show a magnet's magnetic field.

J The short way of writing joules.

joules Energy is measured in joules.

key A set of questions to help us classify things.

key variables Variables that will have a large effect in an investigation.

kilohertz A measurement for pitch or frequency. 1000 hertz.

kilojoules There are 1000 joules in 1 kilojoule.

kJ The short way of writing kilojoules.

lactic acid A substance produced by anaerobic respiration in animals.

large intestine The part of the gut where waste food is stored and water is absorbed.

lateral thinking Thinking in a different direction.

lava Molten rock from deep below the surface of the Earth that reaches the surface through cracks or volcanoes.

light energy Energy transferred by light.

light sensors Equipment used to detect light.

limestone A type of sedimentary rock formed from the shells of sea creatures, which contains calcium carbonate.

line of best fit A line drawn on a graph that shows the overall trend or pattern.

lodestone Iron oxide, a natural magnet.

magma Molten rock found deep below the surface of the Earth.

magnetic Attracted to a magnet.

magnetic field The space around a magnet where it attracts and repels.

magnetic field lines The magnetic field around a magnet. Magnetic field lines run from the north pole of a magnet to its south pole.

magnetic materials Materials that are attracted to a magnet.

magnetic shielding 'Stopping' a magnetic field by putting the magnet inside (but not touching) a box made of magnetic material.

marble A type of metamorphic rock that is produced when limestone is heated under high pressure.

material A solid, liquid or a gas, not empty space. Sound needs matter to travel through.

melting point The temperature at which a solid element becomes a liquid.

metals Materials that are usually solid and shiny when polished. A few are magnetic.

metamorphic rock Rock formed when sedimentary or igneous rocks are changed by intense heat and/or pressure.

MgO The formula representing magnesium oxide.

microbes Another name for microorganisms.

micrometre A unit of measurement $(\mu m) = \frac{1}{1000}$ mm.

microorganism A very small living thing that can only be seen with a microscope.

microphone Equipment that turns sound into an electrical signal.

minerals All rocks are made up of compounds called minerals. Different rocks are made up of different minerals or different mixtures of minerals.

Compounds of calcium, iron, iodine and other elements, that are needed in the diet in small amounts to keep your body healthy, are also called minerals.

mixture A material that contains more than one substance.

molecule A group of atoms joined together.

mosses Small plants that look like a springy cushion. They reproduce from spores and have very small leaves.

N The short way of writing newtons.

narrow-spectrum antibiotics Antibiotics that kill a narrow range of bacteria.

navigate To plan directions to find the way.

negative correlation When two variables move in opposite directions.

newtons Force is measured in newtons.

noise Sound.

non-interlocking Round grains in rocks which do not fit together, as there are gaps between them, are non-interlocking.

non-metals Materials that are usually solids or gases. They have many different appearances.

normal An imaginary line at 90° to a surface.

north pole One end of a magnet. It attracts the south pole of another magnet.

nucleus The part of a cell that controls everything the cell does.

nutrients Useful substances present in foods.

obese People who are very overweight for their height are obese.

object Something you look at using a mirror or lens to form an image.

oesophagus The part of the gut that links the mouth and stomach, also called the gullet.

opaque A material that does not allow light to pass through, but absorbs it, is opaque.

open question A question that can have several possible answers.

organ A group of different tissues that work together to do a job.

organic Food that has been produced without using manufactured chemicals.

oxygen A non-metallic element that is a gas. Oxygen is used in burning and in respiration.

particle An atom or a molecule.

particle model The idea that everything is made up of particles.

passive immunity Immunity given by an injection of ready-made antibodies into the body.

pathogens Organisms that cause disease.

penicillin An antibiotic.

periodic table A table containing all 113 elements, arranged by their properties into groups (columns) and periods (rows).

periods The seven horizontal rows in the periodic table.

permanent magnet A material that stays as a magnet for many years.

physical changes Reversible changes in which no new substances are made, eg melting, boiling.

physical properties The appearance of a material – whether it is a solid, liquid or gas.

physical weathering Breaking down rocks into smaller pieces, without changing them into new substances. Physical weathering can be caused by water, wind and changes in temperature.

pinhole camera The simplest type of camera – a box with a pinhole at one end and a screen at the other.

pitch Sounds can have a high or low pitch. Faster vibrations make higher-pitched sounds than slower vibrations.

placebo 'Medicine' that does not contain any medicine.

poles The two different ends of a magnet.

pond dipping A technique to find out what organisms live in different parts of a pond.

population The number of organisms of a particular species living in a habitat.

porous A substance such as a rock with lots of tiny holes in it is porous.

positive correlation When two variables move in the same direction.

precipitate A solid made when two liquids react.

predation The hunting of a prey animal by another animal (a predator).

predicting To state before something happens what might happen.

primary colour One of three colours of light that humans can see – green, red or blue.

producer A plant, that produces its own food by photosynthesis.

products The new substances that are formed in a chemical reaction.

properties The appearance of a material and the way it reacts.

proteins Nutrients found in foods such as fish, used in your body for growth and repair.

pure A material that contains only one substance.

pyramid of numbers A drawing of the number of organisms at each level of a food chain.

quadrat A wooden frame measuring one metre on all four sides.

radiation The transfer of thermal energy without particles.

random samples Taking samples from different places without choosing the places deliberately.

ratio A way of showing a scale factor. For example, a scale of 1:10 means you have to multiply your number or measurement by 10 to get the real measurement.

ray A thin beam of light.

reactants The substances that take part in a chemical reaction, that change into the products.

recommended daily allowances (RDA) Levels set by the Government that advise about the amounts of different nutrients eaten each day.

reflection/reflects When light or sound bounces off a surface, it is reflected.

refraction/refracts Bending of light when it travels from one material to another, eg air to water or glass to air.

repel Pushing away force between magnets and magnetic materials.

reproduce To make more organisms of the same species.

respiration The process by which plants and animals break down their food to release the chemical energy from it.

respiratory system An organ system that takes oxygen into the blood and gets rid of carbon dioxide.

rickets A disease caused by not eating enough vitamin D in the diet, in which the bones are soft.

rock cycle A cycle that describes how the three rock types change from one to another over millions of years.

roughage Fibre, found in cereals, fruit and vegetables.

sandstone A type of sedimentary rock made up of grains of sand cemented together.

scatter When light is reflected in many directions by a rough surface, it is scattered.

scurvy A disease caused by not eating enough vitamin C in the diet, in which the gums bleed and the skin does not heal.

secondary colour One of three colours of light produced by mixing two primary colours.

sediment Small pieces of rock and dead living things which build up in layers at the bottoms of lakes or seas over millions of years.

sedimentary layers Layers of sediment that have built up over millions of years and become cemented together into rock.

sedimentary rocks A type of rock made up from layers of sediment that have built up over millions of years and become cemented together.

sexually transmitted diseases Diseases that can be caught from sexual intercourse without protection.

shadow Darkness due to an object blocking the light.

shale A type of sedimentary rock made up of very fine grains.

shock wave Vibrations that travel away from an explosion.

slate A type of metamorphic rock that has a layered structure.

small intestine The part of the gut where enzymes and bile are added in alkaline conditions to digest the different substances in food. Absorption also happens here.

solenoid A coil of wire with an electric current running through it that creates a magnetic field.

sound energy Energy transferred by sound.

sound insulation Material that 'soaks up' sound vibrations and stops sound, eg rubber and foam.

source Where something starts or is produced.

south pole One end of a magnet. It attracts the north pole of another magnet.

spectrum The colours in white light – red, orange, yellow, green, blue, indigo, violet.

spores Structures used for reproduction in mosses and ferns.

states of matter The three states of matter are solid, liquid and gas.

stomach The part of the gut where the food is churned up and mixed with enzymes in acidic conditions.

store Keep something for later use.

strain energy Energy stored in a material because the material is being pulled or pushed.

symbol Sign representing an element, eg Fe is the symbol for iron.

temperature The energy per particle measured in degrees Celsius.

temporary magnet A material that only acts as a magnet when it is in a magnetic field.

texture The feel or appearance of a material.

thermal conductor A material that conducts thermal energy well.

thermal (heat) energy Energy transferred from a hot object to a cooler object.

thermal insulator A material that conducts thermal energy poorly.

tissue A group of similar cells that carry out the same job.

topsoil The top layer of soil, made of tiny grains of rock and humus.

transfers Move from one place to another.

translucent A material which both absorbs and transmits light is translucent.

transmitted When light or sound passes through a material, it is transmitted.

transparent A material that allows (transmits) light through is transparent.

trend A general pattern.

unbalanced forces Forces pushing in different directions where one force is bigger than the other. An unbalanced force makes the object move or speed up or slow down.

upthrust The force caused by water pushing up against an object.

vaccinated Injected with dead or inactive microbes to make you immune to a disease before you catch it.

vaccination An injection of dead or inactive microbes into your body to make you immune to a disease before you catch it.

vacuole A bag inside plant cells that contains a liquid which keeps the cell firm.

vacuum A place where there are no particles.

vascular Having veins.

vegetarian A person who does not eat meat.

veins In animals – a blood vessel in which blood flows towards the heart.

In plants – a tube-like structure that carries water, mineral salts and food around the plant.

vibrates Moves rapidly to and fro. Sound is made when something vibrates.

villi (*singular* **villus**) Finger-like structures in the small intestine which increase the area for the absorption of digested food.

viruses Microbes that are smaller than bacteria. They are not made of cells.

vitamin A substance, such as vitamin C, that is needed in the diet in very small amounts to keep your body healthy.

vitamin C A vitamin in fresh fruit and vegetables.

vitamin D A vitamin found in milk and butter and made in your body in sunlight, which gives you strong bones and teeth.

volcano Mountain or hill from which lava erupts out of the Earth's crust.

water A compound of hydrogen and oxygen. Water is the solvent in which all the chemical reactions in your body take place.

weathering Breaking rock down by chemical or physical processes.

weight The force of gravitational attraction on an object, that makes it feel heavy.

white blood cells A vital part of the immune system, these blood cells help fight against microbes.

Index

Note: page numbers in **bold** are for glossary definitions